深度学习
案例精粹

Deep Learning
By Example

[爱尔兰] 艾哈迈德·曼肖伊（Ahmed Menshawy） 著
洪志伟 曹檑 廖钊坡 译

人民邮电出版社
北京

图书在版编目（CIP）数据

深度学习案例精粹 /（爱尔兰）艾哈迈德·曼肖伊（Ahmed Menshawy）著；洪志伟，曹榴，廖钊坡译. -- 北京：人民邮电出版社，2019.9（2023.7重印）
ISBN 978-7-115-50585-9

Ⅰ. ①深… Ⅱ. ①艾… ②洪… ③曹… ④廖… Ⅲ. ①机器学习—案例 Ⅳ. ①TP181

中国版本图书馆CIP数据核字(2019)第082251号

版权声明

Copyright ©2018 Packt Publishing. First published in the English language under the title *Deep Learning By Example*.
All Rights Reserved.

本书由英国 Packt Publishing 公司授权人民邮电出版社出版。未经出版者书面许可，对本书的任何部分不得以任何方式或任何手段复制和传播。
版权所有，侵权必究。

- ◆ 著　　[爱尔兰] 艾哈迈德·曼肖伊（Ahmed Menshawy）
  译　　洪志伟　曹榴　廖钊坡
  责任编辑　张爽
  责任印制　焦志炜
- ◆ 人民邮电出版社出版发行　北京市丰台区成寿寺路11号
  邮编　100164　电子邮件　315@ptpress.com.cn
  网址　http://www.ptpress.com.cn
  北京七彩京通数码快印有限公司印刷
- ◆ 开本：800×1000 1/16
  印张：24.25　　　　　　　　2019年9月第1版
  字数：343千字　　　　　　 2023年7月北京第4次印刷
  著作权合同登记号　图字：01-2017-7973 号

定价：89.00元
读者服务热线：(010)81055410　印装质量热线：(010)81055316
反盗版热线：(010)81055315
广告经营许可证：京东市监广登字20170147号

内容提要

本书主要讲述了深度学习中的重要概念和技术，并展示了如何使用 TensorFlow 实现高级机器学习算法和神经网络。本书首先介绍了数据科学和机器学习中的基本概念，然后讲述如何使用 TensorFlow 训练深度学习模型，以及如何通过训练深度前馈神经网络对数字进行分类，如何通过深度学习架构解决计算机视觉、语言处理、语义分析等方面的实际问题，最后讨论了高级的深度学习模型，如生成对抗网络及其应用。

本书适合数据科学、机器学习以及深度学习方面的专业人士阅读。

作者简介

Ahmed Menshawy 是爱尔兰都柏林三一学院的研究工程师。他在机器学习和自然语言处理领域拥有超过 5 年的工作经验,并拥有计算机科学硕士学位。他曾在埃及开罗阿勒旺大学(Helwan University)计算机科学系做教学助理,负责机器学习和自然语言处理课程,如机器学习、图像处理等,并参与设计了阿拉伯文字到语音的系统。此外,他还是埃及的 IST Networks 工业研发实验室的机器学习专家。

我要感谢那些与我亲近并支持我的人,特别是我的妻子和我的父母。

技术审稿人简介

Md. Rezaul Karim 是德国弗劳恩霍夫应用信息技术研究所（Fraunhofer FIT）的科学家。他还是德国亚琛工业大学的博士研究生。在加入 FIT 之前，他曾担任过 FIT 爱尔兰调查中心（Insight Center）的数据分析研究员。此前，他曾在韩国三星电子公司担任首席工程师。

他在 C++、Java、R、Scala 和 Python 等方面拥有 9 年的研发经验，并曾发表了有关生物信息学、大数据和深度学习的研究论文。他在 Spark、Zeppelin、Hadoop、Keras、Scikit-learn、TensorFlow、DeepLearning4j、MXNet 和 H2O 等领域具有实际工作经验。

Doug Ortiz 是一位经验丰富的企业云、大数据、数据分析和解决方案架构师，负责架构的设计、开发、重新设计和集成企业解决方案。他精通亚马逊 Web 服务（AWS）、Azure、谷歌云、商务智能、Hadoop、Spark、NoSQL 数据库和 SharePoint 等。

译者序

本书的 3 位译者利用业余时间，结合对深度学习领域的了解，经过 3 个月的努力，完成了对全书的翻译。本书的翻译在内容上忠于原文，在表述上又贴近汉语的表达习惯，同时风格也尽量与原著保持一致，以求尽可能地传达作者的原意。在对专业术语的翻译上，我们努力与业界的使用保持统一，使得初学者在阅读完本书后继续查阅其他学习材料时，不会出现术语或者名词矛盾的情况。

尽管已经对深度学习有了一段较长时间的研究，但是本书仍然让我们耳目一新。本书语言简练，和现在流行的机器学习类图书一样，本书没有侧重于揭示深度学习背后的数学理论基础，而是通过问题简述、方法概括以及代码，让读者能够尽快地理解并上手深度学习的算法。对于缺少理论知识的普通开发人员，本书提供了必需的基础知识，读者可以学到几乎所有的要点来探索和理解深度学习是什么。而对于了解深度学习理论知识却缺少实践开发经验的读者，本书提供了大量的案例，使读者能够快速地动手实现深度学习任务。

由于书中给出的代码非常详细，可以直接运行，对于急于将深度学习应用于实际项目的工程师来说，本书可以称为利器。同时，对于那些对深度学习有一定了解的读者，本书也不失为调整自身代码风格并且进一步加深对深度学习理解的宝贵资源。当然，对于有志于在深度学习领域做出一番成就的读者，在熟读本书后，建议进一步阅读更多的学习资料，去探索深度学习背后的数学支撑。

本书的内容涵盖了大多数的主流深度学习任务，包括图像领域的图像识别、目标检测任务，自然语言处理中的词嵌入、情感分析任务，以及无监督学习任务等。本书介绍了几个深度学习中的经典模型，包括卷积网络、循环神经网络和对抗生成网络，当下深度学习的大多数模型都是在这些基础模型上的改进与组合。本书的内容全面，不同领域的从业人员或多或少都能从中获得启发。

本书使用的编程语言是现在非常流行的 Python，代码简单、优雅，易于上手，深度学习框架也是主流的 TensorFlow 框架，想要进一步学习的读者将会有极多的社区资源。当然，随着编程语言的进一步发展，我们也非常期待以后为读者呈现一本基于其他框架（如 Pytorch、Caffe 等）的图书。我们也期待以后能够将更多的、更新的学习任务和模型呈现给读者，如深度强化学习和注意力模型。

感谢翻译过程中所有帮助和指导过我们的人。由于译者水平有限，书中难免会有疏漏，还望读者不吝提出意见和建议。

前言

本书将首先介绍机器学习的基础和如何可视化机器学习的过程，然后使用一些例子来展示传统的机器学习技术，最后讨论深度学习。在本书中，你会开始建立一些机器学习模型，这些模型最后会有助于你学习神经网络模型。通过这些，读者能够熟悉深度学习的一些基础知识，并了解各种以用户友好的方式支持深度学习的工具。本书旨在使普通开发人员能够获得一些深度学习的实践经验。读者能够学到几乎所有的要点来探索和理解深度学习是什么，并且直接动手实现一些深度学习任务。此外，本书将会使用目前最广泛应用的深度学习框架之一——TensorFlow。TensorFlow 在社区的支持度很高，并且在日益增长，这使得它成为构建复杂深度学习应用的理想选择。

本书读者对象

本书是一本深度学习方面的入门书籍，适合那些想要深入了解深度学习并且动手实现它的读者阅读。阅读本书，不要求读者具有机器学习、复杂统计学和线性代数等方面的背景知识。

本书结构

第 1 章解释数据科学或者机器学习的定义，即在事先没有被告知或者编程的情况下，机器从数据集中学习知识的能力。例如，要编写一个可以识别数字的程序非常困难，该程序以手写数字作为输入图像，输出该图像所代表的是 0~9 的哪个数字。同样地，要判断收到的邮件是不是垃圾邮件这样的分类任务也是很困难的。为了解决这样的任务，数据科学家使用一些数据科学、机器学习领域的学习方法和工具，通过给机器一些能够区分每个数字的可解释特征来教会计算机如何自动识别数字。对于垃圾邮件分类问题也是如此，我们能够通过特定的算法教会计算机如何判断一封邮件是不是垃圾邮件，而不是使用传统的正则化表达式并编写数百条规则来对收到的邮件进行分类。

第 2 章讨论线性模型——数据科学领域的基本学习算法。理解线性模型如何工作对于学习数据科学的过程至关重要，因为它是大多数复杂学习算法（包括神经网络）的基本构建模块。

第 3 章介绍模型的复杂性和评估方法。这对构建成功的数据科学系统非常重要。有很多工具可以用来评估和选择模型，本章将讨论一些工具，通过添加更多描述性特征并从现有数据中提取有用信息，这些工具有助于你提高数据的价值。同样地，本章也会讨论与最优数量特征有关的其他工具，并解释为什么"有大量的特征却只有很少量的训练样本/观测值"会是一个问题。

第 4 章概述使用最广泛的深度学习框架之一——TensorFlow。它在社区的支持度很高，而且在日益增长，这对于构建复杂的深度学习应用来说是一个很好的选择。

第 5 章解释 TensorFlow 背后的主要计算概念——计算图模型，并演示如何通过实现线性回归和逻辑回归模型让读者慢慢走上正轨。

第 6 章解释**前馈神经网络**（Feed-forward Neural Network，FNN），FNN 是一种特殊类型的神经网络，其中神经元之间的连接不构成环。因此，它与本书后续将要研究的神经网络中的其他结构（如循环类型神经网络）不同。FNN 是一种广泛使用的架构，而且它是第一个类型最简单的神经网络。本章将会介绍一种典型的 FNN 架构，并且使用 TensorFlow 中的库来实现它。在讲完了这些概念之后，本章会给出一个数字分类的实际例子。例如，给定一组包含手写数字的图像，如何将这些图像分为 10 个（即 0~9）不同的类别？

第 7 章讨论数据科学中的**卷积神经网络**（Convolutional Neural Network，CNN），CNN 是一种特殊的深度学习架构，该架构使用卷积运算从输入图像中提取相关的可解释特征。把很多 CNN 层连接起来作为 FNN，同时通过这种卷积运算来模拟人脑在识别物体时是怎样工作的。每个皮层神经元对受限区域空间（也称为感受野）中的刺激做出反应。特别地，生物医学成像问题有时会难以解决，但在本章中，你将可以看到如何使用 CNN 来识别图像中的这些模式。

第 8 章介绍 CNN 背后的基础知识和直觉/动机，并在可用于物体检测的最流行的数据集之一上进行演示。同样地，你将会看到 CNN 前面的一些层如何提取关于物体的一些基本特征，而最后的卷积层将会提取到更多的语义级特征，这些特征都是从前几层的基本特征中构建而来的。

第 9 章讨论**迁移学习**（Transfer Learning，TL），迁移学习是数据科学中研究的一个问题，主要涉及在解决特定任务时不断获取知识并使用这些已获得的知识来解决另一个不同但相似的任务。本章将展示数据科学领域中一种使用 TL 的现代实践和常见主题。这里的想法是在处理具有较小数据集的领域时，如何从具有非常大的数据集的领域中来获得帮助。最后，本章将重新探讨 CIFAR-10 目标检测示例，并尝试使用 TL 来缩短训练时间和减小性能误差。

第 10 章介绍**循环神经网络**（Recurrent Neural Network，RNN），RNN 是一类广泛应用于自然语言处理（Natural Language Processing，NLP）的深度学习架构。这套架构使我们能够为当前预测提供上下文信息，并且还具有处理任何输入序列中长期依赖性的特定架构。本章将会演示如何构建一个序列到序列的模型，这对于 NLP 中的许多应用都是很有用的。这里将通过建立一个字符级语言模型来阐述这些概念，并且展示模型如何产生和原始输入序列相似的句子。

第 11 章解释了机器学习是一门主要基于统计学和线性代数的科学。由于反向传播算法，应用矩阵运算在大多数机器学习或深度学习架构中都非常普遍。这也是深度学习（机器学习）在总体上只以实数值作为输入的主要原因。事实上，这和很多的应用相矛盾，如机器翻译、情感分析等，它们以文本作为输入。因此，为了在这个应用中使用深度学习，我们需要以深度学习所接受的方式进行处理。本章将会介绍表征学习领域，这是一种从文本中学习实值表示方式的方法，同时保留实际文本的语义。例如，love 的表示方式应该和 adore 的表示方式十分接近，因为它们都会在非常相似的语境中使用。

第 12 章讨论了自然语言处理中一个热门和流行的应用，即所谓的情感分析。现在大多数人通过社交媒体平台来表达他们对某事的看法，所以对于公司甚至政府来说，利用这些海量文本来分析用户对某事的满意度是非常重要的。在本章中，将使用 RNN 来构建一个情感分析解决方案。

第 13 章介绍自编码器网络，它是当下使用最广泛的深度学习架构之一。它主要用于高效解码任务的无监督学习，也可以用于通过学习特定数据集的编码或者表示方式来降维。本章将使用自编码器展示如何通过构建具有相同维度但噪声更小的另一个数据集，来对当前数据集进行去噪。为了在实践中使用这个概念，这里将会从 MNIST 数据集中提取重要的特征，并尝试以此来观察性能是如何显著提高的。

第 14 章讨论**生成对抗网络**（Generative Adversarial Network，GAN）。这是一种由两个彼此对抗的网络（因此有了名字中的对抗）组成的深度神经网络架构。2014 年，Ian

Goodfellow 和包括 Yoshua Benjio 在内的其他研究人员在蒙特利尔大学的一篇论文（见 arxiv 官网）中介绍了 GAN。提到 GAN，Facebook 的 AI 研究总裁 Yann Lecun 称对抗训练是机器学习过去 10 年中最有趣的想法。GAN 潜力巨大，因为它可以学习模仿任何的数据分布。也就是说，在诸如图像、音乐、语言和文本等任何领域可以训练 GAN 以创造与我们类似的世界。从某种意义上说，它们是机器人艺术家，它们的输出（见 nytimes 官网）令人印象深刻同时也使人们受到鼓舞。

第 15 章指出我们能够使用 GAN 来实现非常多有趣的应用。本章将展示 GAN 的另一个有前景的应用，即基于 CelebA 数据库的人脸生成，并且演示如何将 GAN 用于建立那种对于数据集标记不准确或者缺少标记的半监督学习模型。

附录 A 包括鱼类识别示例的所有代码。

充分利用本书

- 在开始阅读之前，告诉读者需要了解的知识，并明确哪些是假定已知的知识。
- 在开始动手操作之前，阅读所需要的所有额外安装说明和相关信息。

下载示例代码文件

读者可以从异步社区（www.epubit.com）上下载本书的示例代码文件。

本书的代码包也托管在 GitHub 网站上面。同样地，其他丰富的图书和视频中的相关代码包也可以从 GitHub 网站获取。

下载彩色图像

我们还提供了一个 PDF 文件，其中包含了本书中所使用的截图/图的彩色图像，读者可以到异步社区中下载。

本书约定

本书中使用了很多约定版式。

代码块设置如下。

```
html, body, #map {
 height: 100%;
 margin: 0;
 padding: 0
}
```

当我们希望提醒读者注意代码块的特定部分时，相关行或条目以粗体显示。

```
[default]
exten => s,1,Dial(Zap/1|30)
exten => s,2,Voicemail(u100)
exten => s,102,Voicemail(b100)
exten => i,1,Voicemail(s0)
```

任何命令行输入或者输出的写法如下。

```
$ mkdir css
$ cd css
```

 表示警告或重要注意事项。

 表示提示信息和技巧。

资源与支持

本书由异步社区出品，社区（https://www.epubit.com/）为您提供相关资源和后续服务。

配套资源

本书提供源代码和彩图文件，请在异步社区本书页面中点击 配套资源 ，跳转到下载界面，按提示进行操作即可。注意，为保证购书读者的权益，该操作会给出相关提示，要求输入提取码进行验证。

提交勘误

作者和编辑尽最大努力来确保书中内容的准确性，但难免会存在疏漏。欢迎您将发现的问题反馈给我们，帮助我们提升图书的质量。

当您发现错误时，请登录异步社区，按书名搜索，进入本书页面，点击"提交勘误"，输入勘误信息，点击"提交"按钮即可。本书的作者和编辑会对您提交的勘误进行审核，确认并接受后，您将获赠异步社区的 100 积分。积分可用于在异步社区兑换优惠券、样书或奖品。

扫码关注本书

扫描下方二维码,您将会在异步社区微信服务号中看到本书信息及相关的服务提示。

与我们联系

我们的联系邮箱是 contact@epubit.com.cn。

如果您对本书有任何疑问或建议,请您发邮件给我们,并请在邮件标题中注明本书书名,以便我们更高效地做出反馈。

如果您有兴趣出版图书、录制教学视频,或者参与图书翻译、技术审校等工作,可以发邮件给我们;有意出版图书的作者也可以到异步社区在线提交投稿(直接访问www.epubit.com/selfpublish/submission 即可)。

如果您是学校、培训机构或企业,想批量购买本书或异步社区出版的其他图书,也可以发邮件给我们。

如果您在网上发现有针对异步社区出品图书的各种形式的盗版行为,包括对图书全部或部分内容的非授权传播,请您将怀疑有侵权行为的链接发邮件给我们。您的这一举动是对作者权益的保护,也是我们持续为您提供有价值的内容的动力之源。

关于异步社区和异步图书

"**异步社区**"是人民邮电出版社旗下IT专业图书社区,致力于出版精品IT技术图书和相关学习产品,为作译者提供优质出版服务。异步社区创办于2015年8月,提供大量精品IT技术图书和电子书,以及高品质技术文章和视频课程。更多详情请访问异步社区官网 https://www.epubit.com。

"**异步图书**"是由异步社区编辑团队策划出版的精品IT专业图书的品牌,依托于人民邮电出版社近30年的计算机图书出版积累和专业编辑团队,相关图书在封面上印有异步图书的LOGO。异步图书的出版领域包括软件开发、大数据、AI、测试、前端、网络技术等。

异步社区

微信服务号

目录

第 1 章　数据科学——鸟瞰全景 1
  1.1　通过示例了解数据科学 2
  1.2　设计数据科学算法的流程 7
    1.2.1　数据预处理 8
    1.2.2　特征选择 8
    1.2.3　模型选择 9
    1.2.4　学习过程 9
    1.2.5　评估模型 9
  1.3　开始学习 10
  1.4　实现鱼类识别/检测模型 12
    1.4.1　知识库/数据集 12
    1.4.2　数据分析预处理 14
    1.4.3　搭建模型 17
  1.5　不同学习类型 22
    1.5.1　监督学习 22
    1.5.2　无监督学习 23
    1.5.3　半监督学习 24
    1.5.4　强化学习 24
  1.6　数据量和行业需求 25
  1.7　总结 25

第 2 章 数据建模实战——"泰坦尼克号"示例 ············ 27

2.1 线性回归模型 ············ 27
2.1.1 原因 ············ 28
2.1.2 广告——一个财务方面的例子 ············ 28

2.2 线性分类模型 ············ 36

2.3 "泰坦尼克号"示例——建立和训练模型 ············ 38
2.3.1 数据处理和可视化 ············ 39
2.3.2 数据分析——监督机器学习 ············ 44

2.4 不同类型的误差解析 ············ 47

2.5 表现(训练集)误差 ············ 47

2.6 泛化/真实误差 ············ 48

2.7 总结 ············ 48

第 3 章 特征工程与模型复杂性——重温"泰坦尼克号"示例 ············ 49

3.1 特征工程 ············ 49
3.1.1 特征工程的类型 ············ 50
3.1.2 重温"泰坦尼克号"示例 ············ 51

3.2 维度灾难 ············ 62

3.3 重温"泰坦尼克号"示例——融会贯通 ············ 64

3.4 偏差-方差分解 ············ 78

3.5 学习可见性 ············ 80

3.6 总结 ············ 80

第 4 章 TensorFlow 入门实战 ············ 82

4.1 安装 TensorFlow ············ 82
4.1.1 在 Ubuntu 16.04 系统上安装 GPU 版的 TensorFlow ············ 83

4.1.2 在 Ubuntu 16.04 系统上安装 CPU 版的 TensorFlow ··············· 86
4.1.3 在 Mac OS X 上安装 CPU 版的 TensorFlow ··················· 88
4.1.4 在 Windows 系统上安装 CPU/GPU 版的 TensorFlow ············· 88
4.2 TensorFlow 运行环境 ··· 89
4.3 计算图 ··· 90
4.4 TensorFlow 中的数据类型、变量、占位符 ····························· 91
    4.4.1 变量 ··· 91
    4.4.2 占位符 ··· 92
    4.4.3 数学运算 ·· 92
4.5 获取 TensorFlow 的输出 ··· 94
4.6 TensorBoard——可视化学习过程 ···································· 95
4.7 总结 ··· 101

第 5 章 TensorFlow 基础示例实战 ·· 102
5.1 神经元的结构 ·· 102
5.2 激活函数 ··· 104
    5.2.1 sigmoid ·· 105
    5.2.2 tanh ·· 105
    5.2.3 ReLU ··· 105
5.3 前馈神经网络 ·· 106
5.4 需要多层网络的原因 ··· 107
    5.4.1 训练 MLP——反向传播算法 ································ 108
    5.4.2 前馈传播 ·· 109
    5.4.3 反向传播和权值更新 ·· 110
5.5 TensorFlow 术语回顾 ·· 110
    5.5.1 使用 Tensorflow 定义多维数组 ································ 112
    5.5.2 为什么使用张量 ··· 114

		5.5.3 变量 ··· 115
		5.5.4 占位符 ··· 116
		5.5.5 操作 ··· 117
5.6	构建与训练线性回归模型 ··· 118
5.7	构建与训练逻辑回归模型 ··· 123
5.8	总结 ·· 130

第 6 章 深度前馈神经网络——实现数字分类 ··································· 131

- **6.1** 隐藏单元与架构设计 ··· 131
- **6.2** MNIST 数据集分析 ·· 133
- **6.3** 数字分类——构建与训练模型 ··· 135
  - 6.3.1 分析数据 ··· 137
  - 6.3.2 构建模型 ··· 140
  - 6.3.3 训练模型 ··· 144
- **6.4** 总结 ·· 148

第 7 章 卷积神经网络 ··· 149

- **7.1** 卷积运算 ·· 149
- **7.2** 动机 ·· 152
- **7.3** CNN 的不同层 ·· 153
  - 7.3.1 输入层 ··· 153
  - 7.3.2 卷积步骤 ··· 154
  - 7.3.3 引入非线性 ··· 155
  - 7.3.4 池化步骤 ··· 156
  - 7.3.5 全连接层 ··· 157
- **7.4** CNN 基础示例——MNIST 手写数字分类 ··· 159

	7.4.1	构建模型	162
	7.4.2	训练模型	167
7.5	总结		174

第 8 章　目标检测——CIFAR-10 示例 175

8.1	目标检测		175
8.2	CIFAR-10 目标图像检测——构建与训练模型		176
	8.2.1	使用软件包	176
	8.2.2	加载 CIFAR-10 数据集	177
	8.2.3	数据分析与预处理	178
	8.2.4	建立网络	183
	8.2.5	训练模型	186
	8.2.6	测试模型	191
8.3	总结		195

第 9 章　目标检测——CNN 迁移学习 196

9.1	迁移学习		196
	9.1.1	迁移学习背后的直觉	197
	9.1.2	传统机器学习与迁移学习之间的不同	198
9.2	CIFAR-10 目标检测——回顾		199
	9.2.1	解决方案大纲	199
	9.2.2	加载和探索 CIFAR-10 数据集	200
	9.2.3	inception 模型迁移值	204
	9.2.4	迁移值分析	207
	9.2.5	模型构建与训练	211
9.3	总结		219

第 10 章 循环神经网络——语言模型 220

10.1 RNN 的直观解释 220
10.1.1 RNN 的架构 221
10.1.2 RNN 的示例 222
10.1.3 梯度消失问题 224
10.1.4 长期依赖问题 225

10.2 LSTM 网络 226

10.3 语言模型的实现 227
10.3.1 生成训练的最小批 230
10.3.2 构建模型 232
10.3.3 训练模型 238

10.4 总结 243

第 11 章 表示学习——实现词嵌入 244

11.1 表示学习简介 244
11.2 Word2Vec 245
11.3 skip-gram 架构的一个实际例子 248
11.4 实现 skip-gram Word2Vec 250
11.4.1 数据分析与预处理 251
11.4.2 构建模型 257
11.4.3 训练模型 259
11.5 总结 264

第 12 章 神经网络在情感分析中的应用 265

12.1 常用的情感分析模型 265
12.1.1 RNN——情感分析背景 267

	12.1.2 梯度爆炸与梯度消失——回顾	269
12.2	情感分析——模型实现	270
	12.2.1 Keras	270
	12.2.2 数据分析与预处理	271
	12.2.3 构建模型	282
	12.2.4 模型训练和结果分析	284
12.3	总结	288

第 13 章 自动编码器——特征提取和降噪 289

13.1	自动编码器简介	289
13.2	自动编码器的示例	290
13.3	自动编码器架构	291
13.4	压缩 MNIST 数据集	292
	13.4.1 MNIST 数据集	292
	13.4.2 构建模型	293
	13.4.3 训练模型	295
13.5	卷积自动编码器	297
	13.5.1 数据集	297
	13.5.2 构建模型	299
	13.5.3 训练模型	301
13.6	降噪自动编码器	304
	13.6.1 构建模型	305
	13.6.2 训练模型	307
13.7	自动编码器的应用	310
	13.7.1 图像着色	310
	13.7.2 更多的应用	311
13.8	总结	311

第 14 章 生成对抗网络 ... 312

14.1 直观介绍 ... 312
14.2 GAN 的简单实现 ... 313
14.2.1 模型输入 ... 315
14.2.2 变量作用域 ... 316
14.2.3 Leaky ReLU ... 316
14.2.4 生成器 ... 317
14.2.5 判别器 ... 318
14.2.6 构建 GAN 网络 ... 319
14.2.7 训练模型 ... 322
14.2.8 从生成器中采样 ... 327
14.3 总结 ... 328

第 15 章 面部生成与标签缺失处理 ... 329

15.1 面部生成 ... 329
15.1.1 获取数据 ... 330
15.1.2 探讨数据集 ... 331
15.1.3 构建模型 ... 332
15.2 用生成对抗网络进行半监督学习 ... 340
15.2.1 直观解释 ... 340
15.2.2 数据分析与预处理 ... 341
15.2.3 构建模型 ... 345
15.3 总结 ... 359

附录 A 实现鱼类识别 ... 360

第 1 章
数据科学——鸟瞰全景

数据科学或机器学习是一个使机器能够在不被告知数据或编程的情况下从数据集中学习知识的过程。例如，编写一个能够将手写数字作为输入图像并根据输入的图像输出值为 0～9 的程序非常困难。这同样适用于将收到的电子邮件分为垃圾邮件或非垃圾邮件的任务。为了解决这些问题，数据科学家使用数据科学或机器学习领域的学习方法和工具，通过向计算机提供一些可以区分一位数字和另一位数字的解释性特征，教会计算机如何自动识别数字。对于垃圾邮件/非垃圾邮件问题也是如此，我们可以通过特定的学习算法教会计算机如何区分垃圾邮件和非垃圾邮件，而不是使用正则表达式并编写数百条规则来对收到的电子邮件进行分类。

对于垃圾邮件过滤程序，你可以通过基于规则的方法对它进行编码，但它不会用于生产中，比如邮件服务器中的程序。所以建立一个学习系统是一个理想的解决方案。

用户可能每天都在使用数据科学应用程序，却不知道它就是数据科学应用程序。例如，某机构可能使用某些系统来检测大家发布的信件的邮政编码，以便自动将它们转发到正确的区域。如果用户使用亚马逊网站，它们通常会推荐用户购买一些东西，亚马逊就是通过了解用户经常搜索或购买哪些东西做到这一点的。

建立一个训练的机器学习算法需要一些基础的历史数据样本，从中学习如何区分不同的例子，并从这些数据中了解一些知识和趋势。之后，训练算法可用于对未知数据进行预测。学习算法将使用原始历史数据，并将尝试从该数据中了解一些知识和趋势。

本章将全面介绍数据科学，包括数据科学如何像一个黑盒子一样工作，以及数据科学家每天面临的挑战。本章具体讨论以下主题。

- 通过示例了解数据科学。
- 设计数据科学算法的程序。
- 开始学习。
- 实现鱼类识别/检测模型。
- 不同的学习类型。
- 数据量和行业需求。

1.1 通过示例了解数据科学

为了说明为特定数据构建学习算法的周期和挑战,我们考虑一个真实的例子。大自然保护协会正在与其他捕鱼公司合作,来共同监测捕捞活动并保护未来的渔业。大自然保护协会希望在未来使用摄像头来扩大这一监控范围。但是因为部署这些摄像头所产生的数据量非常巨大,而且手动处理的成本非常高,所以大自然保护协会想要开发一种学习算法来自动检测不同种类的鱼并分类,以加速视频审查过程。

图 1.1 所示的是保护协会部署的摄像头所拍摄的图像样本,这些图像将用于构建系统。

在这个例子中,我们的目标是分离不同的物种,包括渔船所捕获的金枪鱼、鲨鱼等。作为一个说明性的例子,把这个问题限制在金枪鱼和月鱼两个物种(见图 1.2)中。

图 1.1 大自然保护协会部署的摄像头拍摄到的画面

图 1.2 金枪鱼类(左)和月鱼类(右)

在将问题限制在仅包含两种类型的鱼之后,我们可以从集合中抽取一些随机图像来发现这两种类型之间的一些物理差异。主要考虑以下物理差异。

- **长度**：相对于月鱼，金枪鱼更长。
- **宽度**：月鱼比金枪鱼宽。
- **颜色**：月鱼比金枪鱼更红，金枪鱼往往是蓝色和白色等。

我们可以将这些物理差异作为特征来帮助学习算法（分类器）来区分这两种鱼类。

对象的解释性特征是我们在日常生活中用来区分周围的物体的东西。即使是婴儿也可以使用这些解释性特征来了解周围环境，数据科学也是如此。为了建立一个可以区分不同对象（例如鱼类）的学习模型，我们需要给它一些解释性特征来学习（例如鱼的长度）。为了使模型更加确定并减少错误，我们可以在一定程度上增加对象的解释性特征。

因为这两种鱼类之间存在物理差异，所以这两种不同的鱼类种群有不同的模型或描述。因此，分类任务的最终目标是让分类器学习这些不同的模型，然后将这两种类型中的一种作为输入。分类器将通过选择最适合该图像的模型（金枪鱼模型或月鱼模型）来对它进行分类。

在这种情况下，摄像头收集到的金枪鱼和月鱼图像将作为分类器的知识库。最初，知识库（训练样本）将被标记。所以说，对于每幅图像，我们会预先知道它是金枪鱼还是月鱼。分类器将使用这些训练样本来对不同类型的鱼进行建模，然后我们可以使用训练阶段中已构建的分类器来自动标记在训练阶段中没有见到的未标记的鱼。这种未标记的数据通常称为**未知数据**。整个训练阶段的周期如图 1.3 所示。

下面讨论分类器的训练阶段是如何工作的。

- **预处理**：在这一步中，我们将尝试使用相关的分割方法从图像中分割鱼。
- **特征提取**：在通过去掉背景将鱼从图像中分割出来之后，测量每幅图像的物理差异（长度、宽度、颜色等），最后会得到图 1.4 所示的结果。

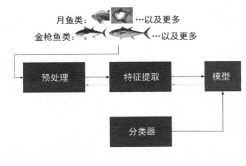

图 1.3　训练阶段周期

我们会将这些数据输入分类器中，以对不同类别的鱼进行建模。

如前所述，用户可以通过之前提出的诸如长度、宽度和颜色这些物理差异（特征）来区分金枪鱼和月鱼。

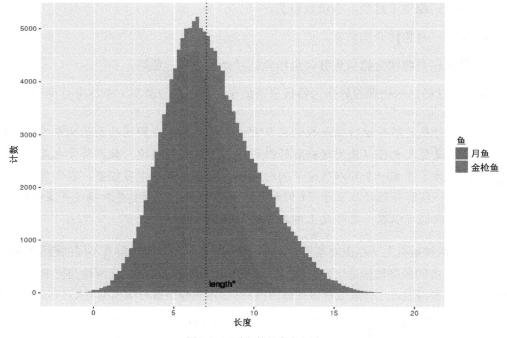

图 1.4 两种鱼的长度直方图

用户也可以利用长度特征来区分两种鱼。因此，我们可以通过观察鱼的长度并判断它是否超过某一个值（length*）来区分不同的鱼类。

所以，基于训练样本，我们可以得出下面的规则。

```
If length(fish)> length* then label(fish) = Tuna
Otherwise label(fish) = Opah
```

为了求 length*，我们可以设法测量训练样本的长度。假定用户测量了样本的长度并得到了图 1.4 所示的直方图。

在这个案例中，我们可以根据长度得出一条规则并用它来区分金枪鱼和月鱼。在这个特定的例子中，length*为 7。所以先前的规则可以更新为：

```
If length(fish)> 7 then label(fish) = Tuna
Otherwise label(fish) = Opah
```

读者可能注意到了，这两个直方图存在重叠，因此这并不是一个很好的结果，这是因为长度特征并不能完全用于区分这两种类型。用户可以考虑纳入诸如宽度等特征，并把它们结合起来。如果我们设法测量训练样本的宽度,可能会得到图 1.5 所示的直方图。

可以看出,只根据一个特征是不能得出准确的结果的,输出模型将会做出很多错误的分类。我们可以设法结合这两个特征,并得到一些合理的模型。

如果结合这两个特征,我们就可能得到图 1.6 所示的散点图。

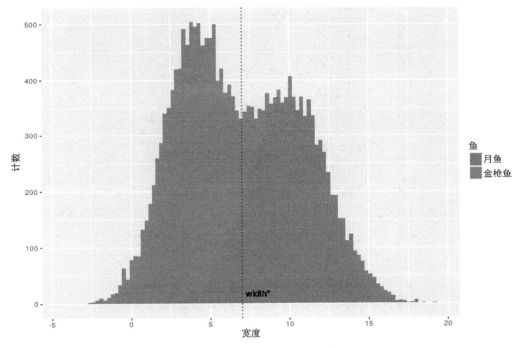

图 1.5 两种鱼的宽度直方图

结合**长度**和**宽度**特征的数值,可以得到图 1.6 所示的散点图。图中浅色点表示金枪鱼,深色点表示月鱼,我们可以认为中间这条黑线是用以区分两种鱼的规则或者决策边界。

举例来说,如果一条鱼的相关数值落在这个决策边界的上方,它就是一条金枪鱼;否则,我们就预测它为一条月鱼。

用户可以设法增加规则的复杂度以避免任何可能的错误,这样做会得到一个新的散点图,如图 1.7 所示。

这个模型的好处是在训练样本上错误分类的数量接近于 0,而实际上这并不是使用数据科学的目的。数据科学的目的是搭建一个能够在未知数据上泛化并表现良好的模型。为了判断该模型是否能够良好地泛化,我们将引入一个称为**测试阶段**的新阶段。

在这个阶段里，用户给训练好的模型一张没有贴标签的图片，希望模型能给这张图片贴上一个正确的标签（**金枪鱼**或者**月鱼**）。

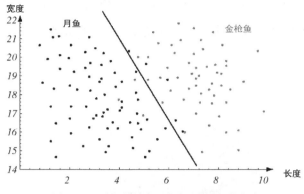

图 1.6　两种鱼的长度和宽度子集之间的组合

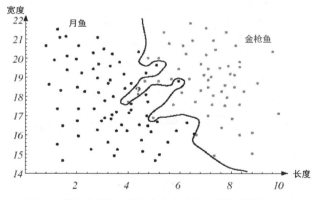

图 1.7　增加决策边界的复杂度来避免训练数据的错误分类

数据科学的最终目的是建立一个能够在生产中（而不是在训练集上）表现良好的模型。所以当我们看到模型在训练集中表现良好时，如图 1.7 所示，不要高兴得太早。大多数情况下，这种模型在识别图像中的鱼种类时表现得并不好。这种模型只在训练数据集上表现良好的状况称为**过拟合**，许多从业者都掉进了这个陷阱。

与其提出一个复杂的模型，不如让用户用一个相对不太复杂的并且能够在测试阶段泛化的模型。图 1.8 所示的是一个不太复杂的模型，该模型可以进一步减少错误分类的数量并且可以很好地泛化未知的数据。

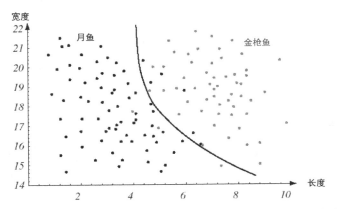

图 1.8 使用一个不太复杂的模型以便于可以在测试样本（未知数据）上泛化

1.2 设计数据科学算法的流程

不同的学习系统通常都遵循相同的设计流程。它们首先获取知识库并从数据中选择相关的可解释特征，然后遍历一系列候选的学习算法并监测每一个算法的性能，最后对它进行评估，以衡量训练过程是否成功。

本节将更详细地叙述所有的设计步骤，如图 1.9 所示。

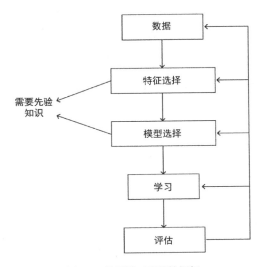

图 1.9 模拟学习流程的框架

1.2.1 数据预处理

学习周期里的数据预处理相当于算法的知识库。为了使学习算法能够对未知数据做出更精确的决策，用户必须以最佳形式提供这种知识库。因此，数据可能需要许多的清洗与预处理（转换）。

1．数据清洗

大多数数据集都需要经历这个步骤，在这个步骤里需要去除数据中的错误、噪声和冗余。数据必须是准确、完整、可靠和无偏的，这是因为使用劣质的知识库可能产生如下许多的问题。

- 结论不精确、有偏。
- 错误率增加。
- 泛化能力降低，泛化能力是指模型在之前训练中未使用过的未知数据上表现良好的能力。

2．数据预处理

在这个步骤里，数据通过一些转换而变得一致且具体。在数据预处理前有许多不同的转换方法可供参考。

- **重命名（重标记）**：这表示将类别的值转换为数值，这是因为类别的值与某些学习算法一起使用是危险的，并且数字会在不同的值之间附加一个顺序。
- **改变尺度（归一化）**：将连续的数值转换或限制在某个范围内，通常是[−1,1]或[0,1]。
- **新特征**：从现有的特征中合成新的特征。比如，肥胖因子=体重/身高。

1.2.2 特征选择

样本的解释特征性（输入变量）可能是非常庞大的，比如用户拿到一个输入 $x_i = (x_i^1, x_i^2, x_i^3, \cdots, x_i^d)$，将它作为训练样本（观察数据/示例），并且 d 非常大。这种情形的一个例子是文档分类任务 3，在这个任务里用户拿到 10 000 个不同的单词，而输入变量则是这些不同单词的出现频次。

这种数量巨大的输入变量可能是有问题的，有时候甚至是灾难性的，因为开发人员有很多的输入变量，却只有很少的训练样本来帮助他们进行学习。为了避免拥有大量输入变量这个问题（维度灾难），数据科学家用降维技术来选取输入变量的一个子集。比如，在文本分类任务中，可以采用以下方法。

- 提取相关的输入（如互信息方法）。
- 主成分分析（PCA）。
- 将相似的单词分组（聚类），这里用到了相似性度量。

1.2.3 模型选择

利用降维技术选择了一个合适的输入变量子集后，接下来就是模型选择了。选择合适的输入变量子集，将使得接下来的学习过程变得非常简单。

在这个步骤里，用户试图找到一个正确的模型来学习。

如果读者之前有过数据科学方面的经验，并曾经将学习方法应用到不同的领域和不同的数据当中，那么这部分内容会相对比较简单，因为这个步骤需要预先了解数据的样貌，以及知道什么样的假设是适合数据的本质的，并基于此来选择合适的学习方法。如果读者没有预备知识也没有关系，因为读者可以通过猜测和尝试不同参数的不同模型，并选择一个在测试集上表现最好的模型，来完成这个步骤。

此外，初始数据分析和数据可视化将会帮助你对数据的分布形式与性质做出更好的猜测。

1.2.4 学习过程

学习指的是用户用来选择最佳模型参数的优化指标。有以下不同的优化指标可以选择：

- **均方误差（MSE）**；
- **最大似然（ML）准则**；
- **最大化后验概率（MAP）**。

有些优化问题可能本身很难解决，但选择正确的模型和误差函数会使得情况有所改善。

1.2.5 评估模型

在这一步中，用户尝试测量模型在未知数据上的泛化误差。既然用户只有特定的数据而无法事先了解任何未知的数据，就可以通过在数据集里随机地选择一部分作为测试集，并且在训练过程中从不使用它，从而使得这部分数据充当未知的有效数据。有很多方法可用于评估所选模型的性能。

- 简单的留出（holdout）法，这种方法将数据简单地分为训练集和测试集。
- 其他更复杂的方法，如基于交叉验证和随机降采样的方法。

这一步的目的是比较在相同数据集上训练出来的不同模型的预测性能，并选择一个拥有最小的测试误差的模型，这个模型将会在未知数据上表现出较好的泛化误差。用户也可以用统计方法判断结果的显著性，以得到泛化误差更准确的信息。

1.3 开始学习

搭建一个机器学习系统将面临一些挑战与问题，本节中我们将尝试解决这些问题。其中的许多问题是在特定领域内存在的，而另一些则不是。

学习的挑战

下面列出了在尝试搭建学习系统时，开发人员通常会遇到的一些挑战和问题。

1. 特征提取——特征工程

特征提取是搭建机器学习系统过程时重要的一步。如果开发人员通过选择适当/正确数量的特征来很好地解决这一挑战，那么学习过程的其余部分将会很简单。此外，特征提取是依赖于领域的，并且需要先验知识来了解哪些特征对于特定任务来说可能是重要的。例如，前文提到的鱼类识别系统的特征与垃圾邮件检测或识别指纹的特征不同。

特征提取将从开发人员拥有的原始数据开始，然后构建衍生的变量/数值（特征），这些特征将为学习任务提供有用的信息，并有助于后续的学习和评估（泛化）。

一些任务具有大量的特征而只有较少的训练样本（观察），用于促进随后的学习和泛化过程。在这种情况下，数据科学家使用降维技术来将大量的特征缩减为一个比较小的集合。

2. 噪声

在鱼类识别任务中，读者可以看到长度、重量、鱼的颜色以及船的颜色可能会有所不同，同时可能还存在阴影、低分辨率的图像以及其他物体。所有这些问题都会对这些本应为鱼类分类任务提供信息的解释性特征产生影响。

一些解决方案在这种情况下会有帮助。例如，有人可能会考虑检测船只的 ID 并且去除船只中某些可能不包含系统所要检测鱼类的部分,这种解决方案将会减小搜索空间。

3. 过拟合

正如你在鱼类识别任务中所看到的,开发人员尝试通过增加模型复杂性并使得模型对训练样本中的每个实例都能完美分类,试图提高模型的性能。正如你在之后将看到的,这类模型并不能在未见过的数据上正常工作(比如开发人员用于测试模型性能的数据)。训练的模型在训练样本集上超常发挥但在测试样本集上表现不佳的情况称为**过拟合**。

如果读者浏览本章的后半部分,那么会看到我们将构建一个学习系统,其目标是将训练样本当作模型的知识库,以便从中学习并在未见过的数据上泛化。训练模型在训练数据集上出现的性能误差并不是用户所感兴趣的;相反,用户比较感兴趣的是训练模型在那些训练过程中未见过的测试样本上出现的误差。

4. 机器学习算法的选择

有时候用户对于模型在某一特定任务上的表现不满意,而需要另一类模型。每种学习策略都有它自己关于数据的假设,而这种假设将成为学习的基础。作为一名数据研究员,读者应该去发掘哪种观点最适用于你的数据,这样读者就有能力去选择尝试哪一类模型而拒绝另一类模型。

5. 先验知识

正如在模型选择和特征提取的概念中所讨论一样,如果读者有以下先验知识,就可以解决这两个问题:

- 合适的特征;
- 模型选择部分。

事先了解鱼类识别系统中的解释性特征使得用户能够区分不同类型的鱼类。通过努力想象分析所拥有的信息,用户可以做出更好的判断并且了解到不同鱼群分类的信息类型。在此先验知识的基础上,就可以选择合适的模型簇了。

6. 缺失值

缺失的特征主要是由于缺少数据或选择了"保密"选项而导致的。用户在学习过程中要怎么处理这类问题呢?例如,某类鱼的宽度特征因为某些原因而丢失了。有很多种方法可以处理这些丢失的特征。

1.4 实现鱼类识别/检测模型

为了介绍机器学习（特别是深度学习）的能力，本节将实现鱼类识别的例子。这并不需要读者了解代码的内部细节。本节重点讲述典型机器学习的流程。

该任务的知识库是一系列图像，其中每张图像都贴上了月鱼或者金枪鱼的标签。在这次实现中，读者将使用其中一种深度学习架构，该架构在图像和计算机视觉领域取得了突破。这种架构叫作**卷积神经网络**（CNN）。卷积神经网络是一类深度学习架构，它通过图像处理的卷积运算来从图像当中提取特征，以解释需要分类的对象。现在，我们可以把它当成一个魔术盒，它将读取图像，从中学习如何区分这两类对象（月鱼和金枪鱼），然后向这个盒子输入未贴标签的图像以测试它的学习过程，并观察它是否能够分辨图像中是哪一类鱼。

 不同类型的学习将在后面的章节中讨论，因此我们将在后面了解为什么上述鱼类识别任务属于监督学习类别。

这个例子中将会使用 Keras。现在我们可以把 Keras 当成一个使得搭建和使用深度学习更加简单的 API。从 Keras 网站上可以看到：

> Keras 是一个高级神经网络 API，用 Python 编写，能够在 TensorFlow、CNTK 或 Theano 之上运行。它旨在实现快速实验，能够以最短的时间把想法转换为结果是做好研究的关键。

1.4.1 知识库/数据集

正如前文所提到的，开发需要以历史数据库作为基础，这将用来使学习算法了解其后期应该完成的任务。但是我们还需要另一个数据集来测试学习算法在学习过程后执行任务的能力。总而言之，在学习过程中需要两类数据集。

第一类数据集是拥有输入数据及其对应标签的知识库，例如鱼的图像和它们的对应标签（月鱼和金枪鱼）。把这类数据输入学习算法中，学习算法从中学习并发现之后将用于分类未标记图像的模式/趋势。

第二类数据集主要用于测试模型将从知识库中学到的知识应用于未标记的图像或未查看的数据上的能力，并检查模型是否运行良好。

正如我们所看到的，开发人员只拥有用来当作学习算法知识库的数据。我们所拥有的所有数据都带有相应的正确的输出。现在需要以某种方法来准备这种不带正确输出的数据（模型将应用于这些数据上）。

在数据科学中，将进行以下操作。

- **训练阶段**：在这个阶段，知识库将提供数据，并把输入数据和正确输出一起提供给模型来训练学习算法/模型。
- **验证/测试阶段**：在这个阶段，我们将度量训练好的模型的效果，还将使用不同的模型属性技术，并通过使用回归的 R^2 分数、分类器的分类错误、IR 模型的召回率和精确率等来度量训练好的模型的性能。

验证/测试阶段通常分为如下两步。

（1）使用不同的学习算法/模型，并基于验证数据选择性能最好的学习算法/模型，此步骤为验证步骤。

（2）度量和报告所选模型在测试集上的准确率，此步骤是测试步骤。

现在我们将学习如何获取这些模型将会用到的数据，并了解它能训练得多好。

既然没有不带正确输出的训练样本，我们就可以从将要使用到的原始训练样本中生成一个。我们可以将数据样本分成 3 个不同的集合（如图 1.10 所示）。

- **训练集**：这个数据集将作为模型的知识库，通常选取原始数据样本中的 70%。
- **验证集**：这个数据集将用来从一系列模型中选择性能最好的模型，通常选取原始数据样本中的 10%。
- **测试集**：这个数据集将用来度量和报告所选模型的准确率，通常它和验证集一样大。

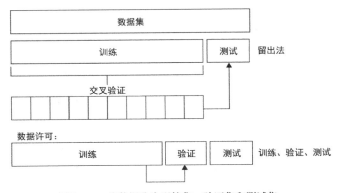

图 1.10 将数据分为训练集、验证集和测试集

这个例子只使用了一种学习算法,因此可以取消验证集,并重新把数据只分为训练集和测试集。通常,数据科学家采用 75/25 的百分比,或者 70/30。

1.4.2 数据分析预处理

本节将对输入图像进行分析和预处理,将它变成这里的卷积神经网络学习算法所接受的格式。

接下来,从导入这个应用所需要的包开始。

```
import numpy as np
np.random.seed(2018)
import os
import glob
import cv2
import datetime
import pandas as pd
import time
import warnings
warings.filterwarnings("ignore")

from sklearn.cross_validation import KFold
from keras.models import Sequential
from keras.layers.core import Dense, Dropout, Flatten
from keras.layers.convolutional import Convolution2D, MaxPooling2D, ZeroPadding2D
from keras.optimizers import SGD
from keras.callbacks import EarlyStopping
from keras.utils import np_utils
from sklearn.metrics import log_loss
from keras import __version__ as keras_version
```

为了使用数据集所提供的图片,我们需要通过以下代码把它们调整为同样大小。从 OpenCV 网站上可以得知,OpenCV 是完成这项工作的一个好选择。

  开源计算机视觉库(Open Source Computer Vision Library,OpenCV)是在 BSD 许可证下发行的,它在学术和商业应用当中都是免费的。它有 C++、C、Python 和 Java 接口,并且支持 Windows、Linux、MacOS、iOS 和 Android 系统。OpenCV 是为高效计算而设计的,并着重于实时应用。OpenCV 用优化的 C/C++

写成，这使得它可以利用多核处理器资源。启用 OpenCL 后，它可以利用底层异构计算平台的硬件加速。

 可以通过使用 Python 软件包管理器和命令 pip install opencv-python 来安装 OpenCV。

```
# Parameters
# ---------------
# img_path : path
#     path of the image to be resized
def resize_image(img_path):
    #reading image file
    img = cv2.imread(img_path)
    #Resize the image to be 32 by 32
    img_resized = cv2.resize(img, (32,32), cv2.INTER_LINEAR)
    return img_resized
```

现在需要加载所有数据集中的训练样本，并根据前面的函数改变每幅图像的大小。下面将实现一个从保存不同鱼类的不同文件夹中加载训练样本的函数。

```
# Loading the training samples and their corresponding labels
def load_training_samples():
    #Variables to hold the training input and output variables
    train_input_variables = []
    train_input_variables_id = []
    train_label = []
    # Scanning all images in each folder of a fish type
    print('Start Reading Train Images')
    folders = ['ALB', 'BET', 'DOL', 'LAG', 'NoF', 'OTHER', 'SHARK', 'YFT']
    for fld in folders:
        folder_index = folders.index(fld)
        print('Load folder {} (Index: {})'.format(fld, folder_index))
        imgs_path = os.path.join('..', 'input', 'train', fld, '*.jpg')
        files = glob.glob(imgs_path)
        for file in files:
            file_base = os.path.basename(file)
            # Resize the image
            resized_img = rezize_image(file)
            # Appending the processed image to the input/output variables of the classifier
            train_input_variables.append(resized_img)
```

```
            train_input_variables_id.append(file_base)
            train_label.append(folder_index)
    return train_input_variables, train_input_variables_id, train_label
```

正如前面所讨论的，有一个测试集将作为未见过的数据来测试模型的泛化能力，所以我们必须对测试图片进行相同的操作：加载它们并改变它们的大小。

```
def load_testing_samples():
    # Scanning images from the test folder
    imgs_path = os.path.join('..', 'input', 'test_stg1', '*.jpg')
    files = sorted(glob.glob(imgs_path))
    # Variables to hold the testing samples
    testing_samples = []
    testing_samples_id = []
    #Processing the images and appending them to the array that we have
    for file in files:
        file_base = os.path.basename(file)
        # Image resizing
        resized_img = rezize_image(file)
        testing_samples.append(resized_img)
        testing_samples_id.append(file_base)
    return testing_samples, testing_samples_id
```

现在需要在另一个函数里调用前一个函数，这个函数将会用 load_training_samples() 函数来加载和改变训练样本的大小。同时，添加几行代码来将训练数据转换为 Numpy 格式，调整数据形状以适应分类器模型，最后把数据转换为浮点数类型。

```
def load_normalize_training_samples():
    # Calling the load function in order to load and resize the training samples
    training_samples, training_label, training_samples_id =load_training_samples()
    # Converting the loaded and resized data into Numpy format
    training_samples = np.array(training_samples, dtype=np.uint8)
    training_label = np.array(training_label, dtype=np.uint8)
    # Reshaping the training samples
    training_samples = training_samples.transpose((0, 3, 1, 2))
    # Converting the training samples and training labels into float format
    training_samples = training_samples.astype('float32')
    training_samples = training_samples / 255
    training_label = np_utils.to_categorical(training_label, 8)
    return training_samples, training_label, training_samples_id
```

对测试集也需要进行相同的操作。

```
def load_normalize_testing_samples():
    # Calling the load function in order to load and resize the testing samples
    testing_samples, testing_samples_id = load_testing_samples()
    # Converting the loaded and resized data into Numpy format
    testing_samples = np.array(testing_samples, dtype=np.uint8)
    # Reshaping the testing samples
    testing_samples = testing_samples.transpose((0, 3, 1, 2))
    # Converting the testing samples into float format
    testing_samples = testing_samples.astype('float32')
    testing_samples = testing_samples / 255
    return testing_samples, testing_samples_id
```

1.4.3 搭建模型

下面开始创建模型了。正如前文提到的，我们将一种名为 CNN 的深度学习架构（如图 1.11 所示）作为鱼类识别任务的学习算法。同样，因为本节只演示在 Keras 和 TensorFlow 深度学习平台的帮助下如何仅使用几行代码来解决复杂的数据科学任务，所以读者不需要了解本章中以前出现过的或即将出现的代码。

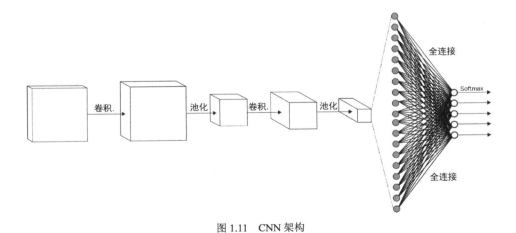

图 1.11 CNN 架构

CNN 和其他深度学习架构的更多详细知识将在后面的章节中介绍。

现在继续构造一个函数以创建鱼类识别任务中使用的 CNN 架构。

```python
def create_cnn_model_arch():
    pool_size = 2 # we will use 2x2 pooling throughout
    conv_depth_1 = 32 # we will initially have 32 kernels per conv.layer...
    conv_depth_2 = 64 # ...switching to 64 after the first pooling layer
    kernel_size = 3 # we will use 3x3 kernels throughout
    drop_prob = 0.5 # dropout in the FC layer with probability 0.5
    hidden_size = 32 # the FC layer will have 512 neurons
    num_classes = 8 # there are 8 fish types
    # Conv [32] ->Conv [32] -> Pool
    cnn_model = Sequential()
    cnn_model.add(ZeroPadding2D((1, 1), input_shape=(3, 32, 32),
        dim_ordering='th'))
    cnn_model.add(Convolution2D(conv_depth_1, kernel_size, kernel_size,
        activation='relu',
dim_ordering='th'))
    cnn_model.add(ZeroPadding2D((1, 1), dim_ordering='th'))
    cnn_model.add(Convolution2D(conv_depth_1, kernel_size, kernel_size,
activation='relu',
        dim_ordering='th'))
    cnn_model.add(MaxPooling2D(pool_size=(pool_size, pool_size),
strides=(2, 2),
        dim_ordering='th'))
    # Conv [64] ->Conv [64] -> Pool
    cnn_model.add(ZeroPadding2D((1, 1), dim_ordering='th'))
    cnn_model.add(Convolution2D(conv_depth_2, kernel_size, kernel_size,
activation='relu',
        dim_ordering='th'))
    cnn_model.add(ZeroPadding2D((1, 1), dim_ordering='th'))
    cnn_model.add(Convolution2D(conv_depth_2, kernel_size, kernel_size,
activation='relu',
        dim_ordering='th'))
    cnn_model.add(MaxPooling2D(pool_size=(pool_size, pool_size),
strides=(2, 2),
        dim_ordering='th'))
    # Now flatten to 1D, apply FC then ReLU (with dropout) and finally softmax (output layer)
    cnn_model.add(Flatten())
    cnn_model.add(Dense(hidden_size, activation='relu'))
    cnn_model.add(Dropout(drop_prob))
    cnn_model.add(Dense(hidden_size, activation='relu'))
    cnn_model.add(Dropout(drop_prob))
    cnn_model.add(Dense(num_classes, activation='softmax'))
```

1.4 实现鱼类识别/检测模型

```
    # initiating the stochastic gradient descent optimiser
    stochastic_gradient_descent = SGD(lr=1e-2, decay=1e-6, momentum=0.9, nesterov=True)    cnn_model.compile(optimizer=stochastic_gradient_descent,
    # using the stochastic gradient descent optimiser
                  loss='categorical_crossentropy')    # using the cross-entropy loss function
    return cnn_model
```

在开始训练模型之前,我们需要使用模型评估和验证方法来帮助评估模型,并检测它的泛化能力。因此,接下来需要用到 **k 折交叉验证法**。读者不需要理解这种方法或它的工作原理,因为本书将在稍后详细解释这种方法。

下面开始并构造一个帮助评估和验证模型的函数。

```
def create_model_with_kfold_cross_validation(nfolds=10):
    batch_size = 16 # in each iteration, we consider 32 training examples at once
    num_epochs = 30 # we iterate 200 times over the entire training set
    random_state = 51 # control the randomness for reproducibility of the results on the same platform
    # Loading and normalizing the training samples prior to feeding it to the created CNN model
    training_samples, training_samples_target, training_samples_id =
       load_normalize_training_samples()
    yfull_train = dict()
    # Providing Training/Testing indices to split data in the training samples
    # which is splitting data into 10 consecutive folds with shuffling
    kf = KFold(len(train_id), n_folds=nfolds, shuffle=True, random_state=random_state)
    fold_number = 0 # Initial value for fold number
    sum_score = 0 # overall score (will be incremented at each iteration)
    trained_models = [] # storing the modeling of each iteration over the folds
    # Getting the training/testing samples based on the generated training/testing indices by
       Kfold
    for train_index, test_index in kf:
        cnn_model = create_cnn_model_arch()
        training_samples_X = training_samples[train_index] # Getting the training input variables
        training_samples_Y = training_samples_target[train_index] # Getting the training output/label variable
        validation_samples_X = training_samples[test_index] # Getting the validation input variables
```

```
        validation_samples_Y = training_samples_target[test_index] # Getting the
validation output/label variable
        fold_number += 1
        print('Fold number {} from {}'.format(fold_number, nfolds))
        callbacks = [
            EarlyStopping(monitor='val_loss', patience=3, verbose=0),
        ]
        # Fitting the CNN model giving the defined settings
        cnn_model.fit(training_samples_X, training_samples_Y,
batch_size=batch_size,
            nb_epoch=num_epochs,
              shuffle=True, verbose=2,
validation_data=(validation_samples_X,
                validation_samples_Y),
              callbacks=callbacks)
        # measuring the generalization ability of the trained model based on
the validation set
        predictions_of_validation_samples =
            cnn_model.predict(validation_samples_X.astype('float32'),
              batch_size=batch_size, verbose=2)
        current_model_score = log_loss(Y_valid,
predictions_of_validation_samples)
        print('Current model score log_loss: ', current_model_score)
        sum_score += current_model_score*len(test_index)
        # Store valid predictions
        for i in range(len(test_index)):
            yfull_train[test_index[i]] =
predictions_of_validation_samples[i]
        # Store the trained model
        trained_models.append(cnn_model)
    # incrementing the sum_score value by the current model calculated score
    overall_score = sum_score/len(training_samples)
    print("Log_loss train independent avg: ", overall_score)
    #Reporting the model loss at this stage
    overall_settings_output_string = 'loss_' + str(overall_score) +
'_folds_' + str(nfolds) +
       '_ep_' + str(num_epochs)
    return overall_settings_output_string, trained_models
```

在构建好了模型并使用k折交叉验证法来评估和验证模型之后,我们需要在测试集上报告训练模型的结果。为了做到这一点,将使用k折交叉验证法,但是这次是在测试集中观察训练模型的效果。

接下来，需要以受过训练的 CNN 模型作为输入的函数，然后使用现有的测试集来测试它：

```
def test_generality_crossValidation_over_test_set(
overall_settings_output_string, cnn_models):
    batch_size = 16 # in each iteration, we consider 32 training examples at once
    fold_number = 0 # fold iterator
    number_of_folds = len(cnn_models) # Creating number of folds based on the
value used in the training step
    yfull_test = [] # variable to hold overall predictions for the test set
    #executing the actual cross validation test process over the test set
    for j in range(number_of_folds):
        model = cnn_models[j]
        fold_number += 1
        print('Fold number {} out of {}'.format(fold_number,
number_of_folds))
        #Loading and normalizing testing samples
        testing_samples, testing_samples_id =
load_normalize_testing_samples()
        #Calling the current model over the current test fold
        test_prediction = model.predict(testing_samples,
batch_size=batch_size, verbose=2)
        yfull_test.append(test_prediction)
    test_result = merge_several_folds_mean(yfull_test, number_of_folds)
    overall_settings_output_string = 'loss_' +
overall_settings_output_string \ + '_folds_' +
        str(number_of_folds)
    format_results_for_types(test_result, testing_samples_id, overall_settings_output_string)
```

1. 模型训练与测试

下面可以通过调用主函数 create_model_with_kfold_cross_validation() 来创建和训练 CNN 模型了，该 CNN 模型使用了 10 折交叉验证法。然后调用测试函数来度量模型在测试集上的泛化能力。

```
if __name__ == '__main__':
    info_string, models = create_model_with_kfold_cross_validation()
    test_generality_crossValidation_over_test_set(info_string, models)
```

2. 鱼类识别——完整的代码

在解释完鱼类识别示例的主要构建模块之后，就可以将所有代码片段连接在一起，并查看如何通过几行代码来构建这样一个复杂的系统。完整的代码放在本书附录 A 中。

1.5 不同学习类型

Arthur Samuel 说："数据科学使得计算机能够在没有明确编程的情况下具备学习能力。"所以任何通过读取训练样例以便在没有明确编程的情况下对未见过的数据进行决策的软件都可以视为学习。数据科学或学习有 3 种不同的形式。

图 1.12 所示为常见的数据科学/机器学习类型。

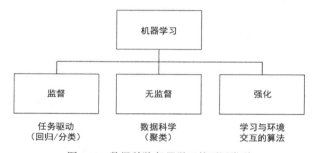

图 1.12 数据科学/机器学习的不同类型

1.5.1 监督学习

大部分数据科学家都使用监督学习。监督学习是指这样的一种情况：具有一些称为输入变量（X）的解释性特征；同时带有与训练样本相关联的标签，这些标签称为输出变量（Y），监督式学习算法的目标是学习从输入变量（X）到输出变量（Y）的映射函数。

$$Y = f(X)$$

因此，监督学习算法将尝试近似地学习从输入变量（X）到输出变量（Y）的映射，以便后续可以使用它来预测未知样本的 Y 值。

图 1.13 所示为监督数据科学系统的典型工作流程。

这种学习称为**监督学习**，因为每个训练样本都有标签/输出。在这种情况下，我们称学习过程受一个监督者监督。算法在训练样本上做决策，然后监督者根据数据的正确标签进行纠正。当监督学习算法达到一个可接受的准确率水平后,学习过程就会终止。

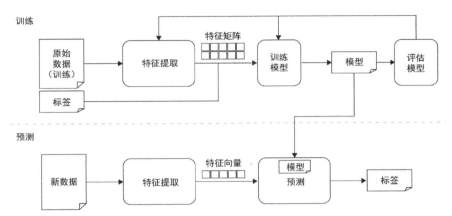

图1.13　一个典型的监督学习工作流程。上面的部分显示了训练过程,从把原始数据输入特征提取模块当中开始,用户将在这里选取有意义的解释性特征来代表数据。然后,将提取/选择的解释特征与训练集相结合,并将它反馈到学习算法以便从中学习。最后做一些模型评估来调整参数,并获得学习算法以从数据样本中获得最好的结果

监督学习有两种形式——分类任务和回归任务。

- **分类任务**:标签或者输出变量是类别的情况,比如月鱼和金枪鱼,垃圾邮件和非垃圾邮件。
- **回归任务**:输出变量是实数的情况,比如房价或者身高。

1.5.2　无监督学习

无监督学习被视为信息研究人员使用的第二种最常见的学习方式。在这种类型的学习中,数据只给出了解释性特征或输入变量(X),而没有任何相应的标签或输出变量。

无监督学习算法的目标是收集信息中隐藏的结构和样例。由于没有与训练样本相关的标记,因此这种学习称为**无监督学习**。无监督学习是一个没有修正的学习过程,它将尝试自行找到基本结构。

无监督学习可以进一步划分为两种形式——聚类任务和关联规则学习任务。

- **聚类任务**:发现训练样本中相似的簇并把它们划分到一组,比如,按照话题划分文档。
- **关联规则学习任务**:发掘一些能够描述训练样本之间关系的规则,比如,观看电

影 X 的人群倾向于观看电影 Y。

图 1.14 所示为一个简单的无监督学习示例,在这个例子中用户有一堆分散的文档,尝试把相似的文档划分在一块。

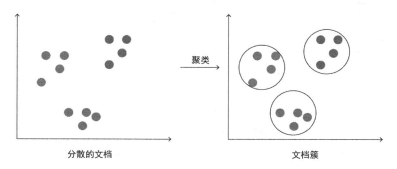

图 1.14　无监督学习利用相似性度量(如欧氏距离)来将相似的文档进行分组并给出对应的决策边界

1.5.3　半监督学习

半监督学习是介于监督学习和无监督学习之间的一种学习类型,用户可以使用输入变量(X)进行训练,但只有其中的一部分转入变量通过输出变量(Y)进行标记/标记。

这类学习的一个很好的例子是 Flickr,Flickr 中有很多由网站用户上传的图片,但只有其中的一小部分图片贴了标签(例如日落、海洋和狗等),其余的都没有标签。

为了解决这类学习任务,开发人员可以从下面两种方法选择一种或者同时使用二者。

- **监督学习**:训练学习算法对未标记的数据做出预测,然后将整个训练样本输入模型中,从中学习并在未知的数据上进行预测。
- **无监督学习**:使用无监督学习算法来学习解释性特征或输入变量的基础结构,就好像没有任何标记的训练样本一样。

1.5.4　强化学习

机器学习的最后一种学习类型是强化学习,在强化学习中只有奖励信号,而没有监督者。

强化学习算法会尝试做出一个决策,然后获得一个奖励信号,这个信号用于指出这个决策是否正确。此外，这种监督反馈或奖励信号可能不会立即出现，而是会有一定的延迟。例如，该算法现在会做出决定，但过了一段时间后，奖励信号才指出了决策是好还是坏。

1.6 数据量和行业需求

数据是学习计算的信息库：任何出色或富有想象力的想法在缺少信息的情况下都将毫无意义。所以如果用户有一个可靠的信息科学应用以及正确的信息，那么就算做好准备工作了。

无论信息结构是什么样的，很明显，如今我们都有能力从信息中挖掘并获取一些激励。然而，因为海量信息已经司空见惯，所以大家都希望信息科学中的仪器和新技术有能力在合理的学习时间内处理这些海量信息。当今任何事情都在产生信息，而是否有能力接受它是一种考验。大型的组织（如谷歌、微软、IBM 等）都在发展自己的海量信息科学规划，以处理由他们的客户每天产生的海量信息。

TensorFlow 是一个开源的机器智能/数据科学平台，在 2016 年 11 月 9 日由谷歌发布。它是一个可扩展的分析平台，使数据科学家能够在可见的时间内构建具有大量数据的复杂系统，并且使他们能够使用贪婪的学习方法，这些学习方法为了获得好的性能需要大量的数据。

1.7 总结

本章讲述了如何构建一个鱼类识别学习系统，并讲述了如何在 TensorFlow 和 Keras 的帮助下使用几行代码来构建复杂的应用程序，如识别鱼类的应用程序。这个代码示例并不要求读者理解，而是为了展示构建的复杂系统的可见性，以及在一般和特定的深度学习中数据科学如何成为一种易于使用的工具。

本章还展示了数据科学家在构建学习系统时可能遇到的一些挑战。

本章还研究了构建学习系统的典型设计周期，并解释了此周期中涉及的每个组件的总体思路。

最后，本章介绍了不同的学习类型，大企业或小公司每天都产生的大数据，大量的数据如何对构建可扩展的工具造成挑战，以及这些工具如何从数据中分析和挖掘到有价值的东西。

  对于目前所提到的这些内容，读者可能会感到有点不知所措。不用担心，本章所讲到的大部分内容都会在其他章节中继续讨论，包括数据科学的挑战以及鱼类识别示例。本章的主要目的是让读者对数据科学及其开发周期有一个总体的了解，但不要求读者对其中的挑战和代码示例有深刻的理解。本章展示了一些代码示例以克服数据科学领域大多数新手的恐惧，并向他们展示如何用几行代码完成鱼识别等复杂系统。

  接下来，本书将通过一个例子来讲解数据科学的基本概念，开启实战之旅。下一部分将主要通过著名的"泰坦尼克"示例来为后面的高阶内容做准备。本书还将讨论许多概念，包括回归和分类的不同学习方法，不同类型的性能错误，哪种性能错误更应该关注，以及一些与解决数据科学挑战和处理不同形式的数据样本相关的问题。

第 2 章 数据建模实战——"泰坦尼克号"示例

线性模型是数据科学领域的基本学习算法。理解线性模型如何工作对于学习数据科学至关重要,因为它是大多数复杂学习算法(包括神经网络)的基本构建模块。

本章将深入讲解数据科学领域的一个著名问题——"泰坦尼克号"示例。介绍这个例子的目的是引入线性模型以进行分类,并让读者看到一个完整的机器学习系统的运行过程,从数据的处理和探索到模型的评估。本章将介绍以下主题。

- 线性回归模型。
- 线性分类模型。
- "泰坦尼克号"示例——建立和训练模型。
- 不同类型的误差解析。

2.1 线性回归模型

线性回归模型是最基本的回归模型,并且广泛用于可预测数据的分析。回归模型的总体思路是检查以下两件事情。

- 一组解释性特征/输入变量是否在预测输出变量方面做得很好?该模型是否使用了可以解释因变量(输出变量)变化的特征?
- 哪些特征是因变量的重要特征?它们以何种方式影响因变量(由参数的大小和符

号表示)？回归参数用于解释一个输出变量(因变量)与一个或多个输入要素(自变量)之间的关系。

回归方程可表达输入变量(自变量)对输出变量(因变量)的影响。具有一个输入变量和一个输出变量的等式是回归方程最简单的形式，回归方程定义为 $y = c+bx$。这里，y 是被预测的因变量，c 是常数，b 是回归参数/系数，x 是输入(自)变量。

2.1.1 原因

线性回归模型是许多学习算法的基础模块，但这并不是它们受欢迎的唯一原因。以下是它受欢迎的关键因素。

- **广泛使用**：线性回归是最古老的回归方法，它广泛应用于许多应用，如预测和财务分析。
- **运行速度快**：线性回归算法非常简单，不包含复杂的数学计算。
- **易于使用（不需要很多调参操作）**：线性回归非常易于使用，而且在大多数情况下，它是读者在机器学习或数据科学课程中学习的第一种学习方法，因为它没有太多的超参数需要调整以获得更好的性能。
- **高度可解释**：因为线性回归简单而且易于检查每个预测-系数对（Predictor-Coefficient Pair）的贡献，所以线性回归是高度可解释的。读者可以很容易地理解模型行为，并为非技术人员解释模型输出。如果系数为零，则相关的预测变量不起任何作用。如果系数不为零，则可以很容易地确定特定预测变量做出的贡献。
- **是许多其他方法的基础**：线性回归是许多学习方法的基础，如神经网络及其加强版、深度学习。

2.1.2 广告——一个财务方面的例子

为了更好地理解线性回归模型，本节将介绍一个有关广告的例子。我们将尝试通过研究不同公司在电视、广播和报纸上花费的广告金额来预测公司的销售情况。

1. 相关性

为了用线性回归模型来建模本节中的广告数据样本，我们使用统计模型库来获得线性模型的良好特性。稍后，我们将使用 scikit-learn，它对于一般的数据科学非常有用。

2. 使用 pandas 导入数据

Python 中有很多库可用于读取、转换或写入数据，pandas 就是其中的一个库。pandas 是一个开源的库，它具有强大的数据分析功能工具以及非常简单的数据结构。

读者可以通过许多不同的方式轻松获得 pandas，其中最好的方法是通过 conda 安装它。

"conda 是一个开源软件包管理系统和环境管理系统，用于安装多个版本的软件包及其依赖关系并在它们之间轻松切换。它适用于 Linux 系统、Mac OS X 和 Windows 系统。虽然它是为 Python 程序创建的，但它可以打包并分发任何语言的软件。"

——conda 官网

我们可以通过安装 Anaconda 轻松获得 conda，Anaconda 是一个开放的数据科学平台。

接下来，我们来看看如何使用 pandas 来读取广告数据样本。首先，读者需要导入 pandas。

```
import pandas as pd
```

接下来，可以使用 pandas.read_csv 方法将数据加载到一个名为 **DataFrame** 的易于使用的 pandas 数据结构中。有关 pandas.read_csv 及其参数的更多信息，可以参考此方法的 pandas 文档。

```
# read advertising data samples into a DataFrame
advertising_data = pd.read_csv('http://www-bcf.usc.edu/~gareth/ISL/Advertising.csv', index_col=0)
```

传递给 pandas.read_csv 方法的第一个参数是表示文件路径的字符串值。该字符串可以是 http、ftp、s3 和文件的 URL。传递的第二个参数是将用作数据行的标签/名称的列的索引。

现在，我们有 DataFrame 数据，其中包含 URL 中提供的广告数据，每行的标记是第一列的数值。如前所述，pandas 提供易于使用的数据结构，我们可以将它当作数据容器。这些数据结构带有对应的一些方法，我们可以使用这些方法来转换或操作自己的数据。

现在，看一下广告数据的前 5 行，代码如下。

```
# DataFrame.head method shows the first n rows of the data where the
# default value of n is 5, DataFrame.head(n=5)
advertising_data.head()
```

输出如下。

	TV	Radio	Newspaper	Sales
1	230.1	37.8	69.2	22.1
2	44.5	39.3	45.1	10.4
3	17.2	45.9	69.3	9.3
4	151.5	41.3	58.5	18.5
5	180.8	10.8	58.4	12.9

3．理解广告数据

这个问题属于监督学习类型的问题，其中有解释性特征（输入变量）和输出（输出变量）。输入变量如下。

- **电视投入**：在特定市场上为单一产品花费在电视上的广告费用（单位是千美元）。
- **广播投入**：花费在广播上的广告费用。
- **报纸**：花费在报纸上的广告费用。

输出/结果/输出变量是**销量**，即在特定市场中单件产品的销量（单位是千件）。

可以使用 DataFrame 中的 shape 方法来了解数据中样本的形状。

```
# print the shape of the DataFrame
advertising_data.shape
Output:
(200, 4)
```

所以可以得知数据中共有 200 个样本。

4．数据分析和可视化

为了理解数据的基本形式、特征与输出之间的关系以及其他信息，我们可以使用不同类型的可视化图。为了理解广告数据特征和输出之间的关系，本节将使用散点图。

为了对数据进行不同类型的可视化，我们可以使用 Matplotlib，这是一个用于可视化的 Python 2D 库。要获得 Matplotlib，可以按照 Matplotlib 官网中的安装说明进行操作。

下面导入可视化库 Matplotlib。

```
import matplotlib.pyplot as plt

# The next line will allow us to make inline plots that could appear
directly in the notebook
# without poping up in a different window
%matplotlib inline
```

现在，使用散点图来可视化广告数据特征和输出变量之间的关系。

```
fig, axs = plt.subplots(1, 3, sharey=True)

# Adding the scatterplots to the grid
advertising_data.plot(kind='scatter', x='TV', y='sales', ax=axs[0],
figsize=(16, 8))
advertising_data.plot(kind='scatter', x='radio', y='sales', ax=axs[1])
advertising_data.plot(kind='scatter', x='newspaper', y='sales', ax=axs[2])
```

输出如图 2.1 所示。

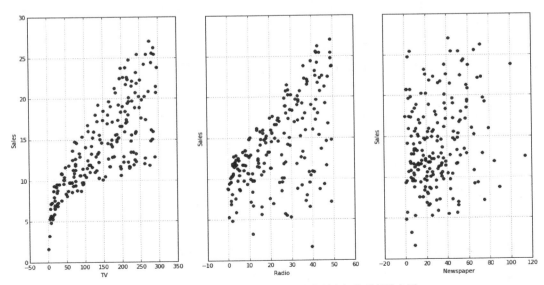

图 2.1　用于理解广告数据特征与响应变量之间关系的散点图

现在，我们需要了解广告是如何有助于增加销量的。首先，问自己几个问题。值得提出的问题有广告与销量之间的关系，哪种广告对销量的贡献更大，每种广告对销量的影响等。接下来，我们将尝试使用简单的线性模型来回答这些问题。

5．简单回归模型

线性回归模型是一种学习算法，它使用一些**解释性特征**（**输入**或**预测变量**）来预测**定量**（也称为**数值**）**输出**。

只有一个特征的简单线性回归模型采用以下形式。

$$y = \beta_0 + \beta_1 x$$

其中，y 表示预测值（输出），即销量；x 表示输入特征；β_0 表示截距；β_1 表示特征 x 的系数。

β_0 和 β_1 都是模型**系数**。为了创建一个在广告例子中可以预测销量的模型，我们需要学习这些系数。β_1 是特征 x 对输出 y 的学习效果。例如，若 $\beta_1 = 0.04$，则意味着在电视广告上额外花费的 100 美元与增加 4 件销量相关联。接下来，我们看看如何学习这些系数。

（1）学习模型参数

为了评估模型的系数，这里需要用一条回归线来拟合数据，该回归线可以给出和实际销量类似的答案。为了得到最拟合数据的回归线，这里使用名为**最小二乘**的标准。我们需要找到一条线，以最小化预测值与观察值（实际值）之间的差异。换句话说，需要找到一条回归线，该回归线使**残差平方和**（sum of squared residuals）最小，如图 2.2 所示。

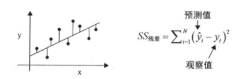

（预测值和观察值之间的差异）

图 2.2　用回归线拟合数据点（电视广告的样本），使残差平方和最小

以下是图 2.2 中存在的一些量。

- **黑点**表示 x（电视广告）和 y（销量）的观察值。
- **虚线**表示最小二乘线（回归线）。
- **实线**表示残差，即预测值与观察值（实际值）之间的差异。

以下是系数与最小二乘线（回归线）的关系。

- β_0 是截距，当 $x = 0$ 时，它等于 y 的值。
- β_1 是斜率，它等于 y 的变化量除以 x 的变化量。

图 2.3 给出了图形化的解释。

现在，继续使用 **Statsmodels** 来学习这些系数。

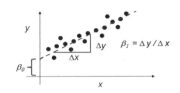

图 2.3 最小二乘线与模型系数之间的关系

```
# To use the formula notation below, we need to import the module like the following
import statsmodels.formula.api as smf
# create a fitted model in one line of code(which will represent the least squares line)
lm = smf.ols(formula='sales ~ TV', data=advertising_data).fit()
# show the trained model coefficients
lm.params
```

输出如下。

```
Intercept    7.032594
TV           0.047537
dtype: float64
```

正如之前所提到的，线性回归模型的一个优点是易于解释，因此接下来继续解释模型。

（2）解释模型系数

下面解释模型系数，如电视广告系数（β_1）。输入特征（电视广告）支出增加 1 个单位与销量（输出）增加 0.047 537 个单位有关。换句话说，另外花费在电视广告上的 100 美元与销量中增加的 4.7537 件相关。

用电视广告数据构建学习模型的目标是预测未知数据对应的销量。因此，接下来将讲解基于已知的电视广告支出如何使用学习模型来预测未知的销量。

（3）使用模型来预测

假设有一些之前未见过的电视广告支出数据，现在我们想知道它们对公司销量的影响。我们使用学习模型可以完成这个任务，假如我们想知道 5 万美元的电视广告能增加多少销量。

这里用学习的模型系数做出这样一个计算:
$$y = 7.032\ 594 + 0.047\ 537 \times 50$$

执行以下代码。

```
# manually calculating the increase in the sales based on $50k
7.032594 + 0.047537*50000
```

输出结果如下。

```
9409.444
```

这里也可以使用 Statsmodels 来做预测。首先,需要在 pandas DataFrame 中输入电视广告费用,因为它是 Statsmodels 接口的输入参数。

```
# creating a Pandas DataFrame to match Statsmodels interface expectations
new_TVAdSpending = pd.DataFrame({'TV': [50000]})
new_TVAdSpending.head()
```

输出结果如下。

	TV
0	50000

现在,可以继续使用预测函数来预测销量。

```
# use the model to make predictions on a new value
preds = lm.predict(new_TVAdSpending)
```

输出结果如下。

```
array([ 9.40942557])
```

下面绘制学习好的最小二乘线。绘制一条线需要两个点,每个点表示为(x, predict_value_of_x)。

因此,计算电视广告费用的最小值和最大值。

```
# create a DataFrame with the minimum and maximum values of TV
X_min_max = pd.DataFrame({'TV': [advertising_data.TV.min(),
advertising_data.TV.max()]})
X_min_max.head()
```

输出结果如下。

	TV
0	0.7
1	296.4

分别计算这两个值的预测值。

```
# predictions for X min and max values
predictions = lm.predict(X_min_max)
predictions
```

输出结果如下。

```
array([7.0658692, 21.12245377])
```

现在，我们绘制实际数据，然后用最小二乘法线进行拟合。

```
# plotting the acutal observed data
advertising_data.plot(kind='scatter', x='TV', y='sales')
#plotting the least squares line
plt.plot(new_TVAdSpending, preds, c='red', linewidth=2)
```

输出结果如图 2.4 所示。

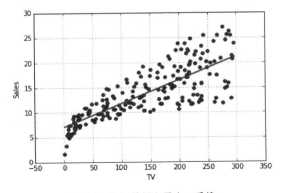

图 2.4 实际数据和最小二乘线

这个例子的扩展和进一步的解释将在第 3 章中详细介绍。

2.2 线性分类模型

本节将解释逻辑回归,它是广泛使用的分类算法之一。

什么是逻辑回归?逻辑回归的简单定义是它是一种涉及线性判别式的分类算法。

下面将用两点阐明这一定义。

- 与线性回归不同,逻辑回归不会估计给定一组特征或输入变量的数值变量的值。相反,逻辑回归算法的输出是给定样本观察值属于特定类别的概率。假设有一个二元分类问题,在这种类型的问题中,输出变量只有两类,例如患病或未患病。因此,若某个样本属于患病类别的概率是 P_0,那么某个样本属于未患病类别的概率是 $P_1=1-P_0$。因此,逻辑回归算法的输出总是介于 0 和 1。

- 正如读者可能知道的那样,用于回归或分类的学习算法有很多,并且每种学习算法本身都有关于数据样本的假设。选择适合数据的学习算法的能力将随着实践和对该主题的深入理解而逐渐形成。逻辑回归算法的中心假设是输入(特征)空间可以由一个线性平面分成两个区域(每个类别一个区域),如果有两个特征,那么这个线性平面是一条分界线;如果有 3 个特征,那么这个线性平面就是分界面,依次类推。这个边界的位置和方向将由具体的数据决定。如果数据满足线性可分这个约束条件,就可以用线性平面将每个类划分到相对应的区域中。图 2.5 说明了这个假设。在图 2.5 中,输入(特征)有 3 个维度,以及两个可能的类别:患病(深灰色)和非患病(黑色)。将两个区域相互分开的分界面称为**线性判别式**,这是因为它是线性的且有助于模型区分属于不同类别的样本。

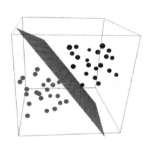

图 2.5 划分两个类别的线性分界面

如果数据样本不是线性可分的,那么可以通过添加更多的输入特征来将数据转换到更高维的空间来实现。

分类与逻辑回归

我们已经学习了如何将连续量(例如,电视广告费用对公司销量的影响)预测为输入值(例如,电视、广播和报纸广告方面的费用)的线性函数。但对于其他任务,输出并不会是连续量。例如，预测某人是否患病是一种分类问题，我们需要另一种学习算法来执行此操作。本节将深入研究逻辑回归的数学分析，这是一种用于分类任务的学习算法。

在线性回归中，我们尝试使用线性模型函数 $y=h_\theta(x)=\theta^T x$ 来预测该数据集中第 i 个样本 $x^{(i)}$ 的输出变量 $y^{(i)}$ 的值。但对于预测二元标签($y^{(i)}\in\{0,1\}$)等分类任务来说，这并不是一个很好的解决方案。

逻辑回归是用于分类任务的众多学习算法之一，使用不同的假设类，我们可以预测特定样本属于第 1 个类的概率以及它属于第 0 个类的概率。因此，在逻辑回归中，我们将尝试学习以下函数。

$$P(y=1|x)=h_\theta(x)=\frac{1}{1+\exp(-\theta^T x)}\equiv\sigma(\theta^T x)$$
$$P(y=0|x)=1-P(y=1|x)=1-h_\theta(x)$$

函数 $\sigma(z)=\dfrac{1}{1+\exp(-z)}$ 通常称为 **sigmoid** 函数或 **logistic** 函数，它将 $\theta^T x$ 的值压缩到固定范围[0,1]，如图 2.6 所示。因为把值压缩在[0,1]，所以可以将 $h_\theta(x)$ 解释为概率。

读者的目标是求 θ 的值，使得当输入样本 x 属于第 1 类时，概率 $P(y=1|x)=h_\theta(x)$ 比较大；当 x 属于第 0 类时，概率 $P(y=0|x)=h_\theta(x)$ 较大。

因此，假设有一组训练样本及其对应的二分类标签 $\{(x^{(i)},y^{(i)}):i=1,\cdots,m\}$，读者需要做的是最小化下面的代价函数，它用于衡量给定 h_θ 的好坏。

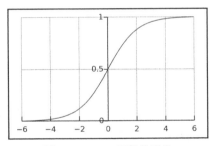

图 2.6 sigmoid 函数的形状

$$J(\theta)=-\sum_i(y^{(i)}\log_2(h_\theta(x^{(i)}))+(1-y^{(i)})\log_2(1-h_\theta(x^{(i)})))$$

注意，对于每个训练样本，方程的两个项中，只有一项非零（根据标签 $y^{(i)}$ 的值是否为 0）。当 $y^{(i)}=1$ 时，最小化模型代价函数需要使 $h_\theta(x^{(i)})$ 尽量大；当 $y^{(i)}=0$ 时，最小化模型代价函数需要使 $1-h_\theta(x^{(i)})$ 尽量大。

现在，用一个代价函数来计算给定的假设 h_θ 拟合训练样本的程度。我们可以使用优化法来最小化 $J(\theta)$，并求参数 θ 的最优解来学习如何对训练样本进行分类。完成此操作后，

我们就可以使用这些参数判断这两个类标签（0 和 1）中哪一个最有可能，从而对新样本进行分类。如果 $P(y=1|x) < P(y=0|x)$，那么输出 0；否则，输出 1。同理，如果阈值定义为 0.5，当 $h_\theta(x) > 0.5$ 时输出 1；否则，输出 0。

为了最小化代价函数 $J(\theta)$，我们可以使用优化法来求出使代价函数最小化的 θ 的最佳值。因此，我们可以使用名为**梯度**的微积分工具，该工具试图求出成本函数的最大增长率。然后，可以从相反的方向来求出这个函数的最小值。例如，$J(\theta)$ 的梯度由 $\nabla_\theta J(\theta)$ 表示，这表示取相对于模型参数的代价函数的梯度。因此，需要提供一个函数来计算任意 θ 的 $J(\theta)$ 和 $\nabla_\theta J(\theta)$。如果推导以上的代价函数 $J(\theta)$ 相对于 θ_j 的梯度或导数，将得到以下结果。

$$\frac{\partial J(\theta)}{\partial \theta_j} = \sum_i x_j^{(i)} (h_\theta(x^{(i)}) - y^{(i)})$$

将它写成向量形式，如下所示。

$$\nabla_\theta J(\theta) = \sum_i x^{(i)} (h_\theta(x^{(i)}) - y^{(i)})$$

现在，我们从数学上了解了逻辑回归，接下来继续使用这种新的学习方法来解决分类任务。

2.3 "泰坦尼克号"示例——建立和训练模型

"泰坦尼克号"沉船是历史上最悲惨的事件之一。这起事件导致 2224 名乘客和 1502 名船员死亡。在本节中，我们将使用数据科学来预测乘客是否能在这场悲剧中生存，然后根据这场悲剧的实际统计数据来测试这个模型的性能。

要开始学习"泰坦尼克号"示例，我们需要做以下工作。

1）从 GitHub 网站下载 ML_Titanic ZIP 格式的存储库或者在命令行窗口中执行命令 Git clone 命令复制 GitHub 网站中的 ML_Titanic.git。

2）安装 virtualenv。

3）导航到解压缩或复制存储库的目录，并使用 virtualenv ml_titanic 创建虚拟环境。

4）使用 source ml_titanic/bin/activate 激活环境。

5）安装依赖的包 pip install -r requirements.txt。

6）在命令行窗口或者终端中执行 ipython notebook。

7）按照本章中的示例代码进行操作。

8）当结束操作后，使用 deactivate 关闭虚拟环境。

2.3.1 数据处理和可视化

在本节中，我们将要做一些数据预处理和分析。可以认为数据预处理和分析是机器学习中最重要的步骤之一，也可以认为它是最重要的步骤，没有之一。因为在这一步，我们将了解在训练过程一直与我们相伴的数据。此外，了解使用的数据后，我们可以缩小自己的候选算法集，从而选择最优的算法。

首先，导入需要软件包。

```
import matplotlib.pyplot as plt
%matplotlib inline

from statsmodels.nonparametric.kde import KDEUnivariate
from statsmodels.nonparametric import smoothers_lowess
from pandas import Series, DataFrame
from patsy import dmatrices
from sklearn import datasets, svms

import numpy as np
import pandas as pd
import statsmodels.api as sm

from scipy import stats
stats.chisqprob = lambda chisq, df: stats.chi2.sf(chisq, df)
```

用 pandas 读取"泰坦尼克号"中乘客和船员的数据。

```
titanic_data = pd.read_csv("data/titanic_train.csv")
```

接下来，检查数据集的维度，查看样本的数量和描述数据集的解释性特征的个数。

```
titanic_data.shape

Output:
(891, 12)
```

共有 891 个数据样本以及 12 个解释性特征来描述此记录。

```
list(titanic_data)

Output:
['PassengerId',
 'Survived',
 'Pclass',
 'Name',
 'Sex',
 'Age',
 'SibSp',
 'Parch',
 'Ticket',
 'Fare',
 'Cabin',
 'Embarked']
```

接下来，查看样本的详细信息。

```
titanic_data[500:510]
```

输出结果如图 2.7 所示。

	PassengerId	Survived	Pclass	Name	Sex	Age	SibSp	Parch	Ticket	Fare	Cabin	Embarked
500	501	0	3	Calic, Mr. Petar	male	17.0	0	0	315086	8.6625	NaN	S
501	502	0	3	Canavan, Miss. Mary	female	21.0	0	0	364846	7.7500	NaN	Q
502	503	0	3	O'Sullivan, Miss. Bridget Mary	female	NaN	0	0	330909	7.6292	NaN	Q
503	504	0	3	Laitinen, Miss. Kristina Sofia	female	37.0	0	0	4135	9.5875	NaN	S
504	505	1	1	Maioni, Miss. Roberta	female	16.0	0	0	110152	86.5000	B79	S
505	506	0	1	Penasco y Castellana, Mr. Victor de Satode	male	18.0	1	0	PC 17758	108.9000	C65	C
506	507	1	2	Quick, Mrs. Frederick Charles (Jane Richards)	female	33.0	0	2	26360	26.0000	NaN	S
507	508	1	1	Bradley, Mr. George ("George Arthur Brayton")	male	NaN	0	0	111427	26.5500	NaN	S
508	509	0	3	Olsen, Mr. Henry Margido	male	28.0	0	0	C 4001	22.5250	NaN	S
509	510	1	3	Lang, Mr. Fang	male	26.0	0	0	1601	56.4958	NaN	S

图 2.7 "泰坦尼克号"中的样本

现在，我们有一个 Pandas DataFrame，它包含需要分析的 891 名乘客的信息。DataFrame 的列表示每个乘客/船员的解释性特征，例如姓名、性别或年龄。

其中一些解释性功能是完整的，没有任何缺失值，例如幸存特征，它有 891 个条目。其他解释性特征包含缺失值，例如年龄特征，它只有 714 个条目。DataFrame 中的任何缺失值都表示为 NaN。

2.3 "泰坦尼克号"示例——建立和训练模型

如果我们观察所有的数据集特征,会发现 Ticket 和 Cabin 特征有许多缺失值(NaN),因此它们不会为我们的分析贡献很多力量。为了解决这个问题,我们将从 DataFrame 中删除它们。

使用以下代码从 DataFrame 中删除 Ticket 和 Cabin 特征。

```
titanic_data = titanic_data.drop(['Ticket','Cabin'], axis=1)
```

因为某些原因,数据集中会包含缺失值,但为了保持数据集的完整性,需要处理这些缺失值。在这个特定的问题中,这次我们选择删除它们。

使用以下代码来删除所有其他特征中的所有 NaN 值。

```
titanic_data = titanic_data.dropna()
```

现在,我们有了一个完善的数据集,可以用它来进行数据分析了。如果尝试在不删除 Ticket 和 Cabin 特征列表的情况下删除所有 NaN,会发现大部分数据集都会被删除,因为.dropna()方法会从 DataFrame 中删除所有满足条件的,即使某一个特征中只有一个 NaN。

下面可视化数据来查看一些特征的分布并理解解释性特征之间的关系。

```
# declaring graph parameters
fig = plt.figure(figsize=(18,6))
alpha=alpha_scatterplot = 0.3
alpha_bar_chart = 0.55
# Defining a grid of subplots to contain all the figures
ax1 = plt.subplot2grid((2,3),(0,0))
# Add the first bar plot which represents the count of people who survived vs
not survived.
titanic_data.Survived.value_counts().plot(kind='bar', alpha=alpha_bar_chart)
# Adding margins to the plot
ax1.set_xlim(-1, 2)
# Adding bar plot title
plt.title("Distribution of Survival, (1 = Survived)")
plt.subplot2grid((2,3),(0,1))
plt.scatter(titanic_data.Survived, titanic_data.Age, alpha=alpha_scatterplot)
# Setting the value of the y label (age)
plt.ylabel("Age")
# formatting the grid
plt.grid(b=True, which='major', axis='y')
plt.title("Survival by Age, (1 = Survived)")
ax3 = plt.subplot2grid((2,3),(0,2))
```

```
titanic_data.Pclass.value_counts().plot(kind="barh", alpha=alpha_bar_chart)
ax3.set_ylim(-1, len(titanic_data.Pclass.value_counts()))
plt.title("Class Distribution")
plt.subplot2grid((2,3),(1,0), colspan=2)
# plotting kernel density estimate of the subse of the 1st class passenger's age
titanic_data.Age[titanic_data.Pclass == 1].plot(kind='kde')
titanic_data.Age[titanic_data.Pclass == 2].plot(kind='kde')
titanic_data.Age[titanic_data.Pclass == 3].plot(kind='kde')
# Adding x label (age) to the plot
plt.xlabel("Age")
plt.title("Age Distribution within classes")
# Add legend to the plot.
plt.legend(('1st Class', '2nd Class','3rd Class'),loc='best')
ax5 = plt.subplot2grid((2,3),(1,2))
titanic_data.Embarked.value_counts().plot(kind='bar', alpha=alpha_bar_chart)
ax5.set_xlim(-1, len(titanic_data.Embarked.value_counts()))
plt.title("Passengers per boarding location")
```

样本可视化结果如图 2.8 所示。

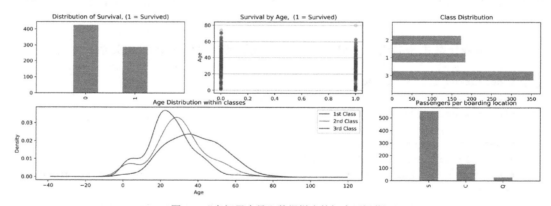

图 2.8 "泰坦尼克号"数据样本的初步可视化

正如前面所提到的，这种分析的目的是根据可用的特征（如 pclass、sex、age 和 fare price）来预测特定乘客是否能够在这次事故中生存。因此，弄清楚我们是否能够更好地了解那些幸存和死亡的乘客。

首先，绘制一个条形图，以查看每个类别（幸存/死亡）中的观察数量。

```
plt.figure(figsize=(6,4))
fig, ax = plt.subplots()
titanic_data.Survived.value_counts().plot(kind='barh', color="blue", alpha=.65)
```

2.3 "泰坦尼克号"示例——建立和训练模型

```
ax.set_ylim(-1, len(titanic_data.Survived.value_counts()))
plt.title("Breakdown of survivals(0 = Died, 1 = Survived)")
```

输出结果如图 2.9 所示。

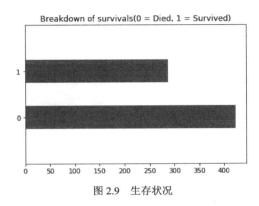

图 2.9 生存状况

通过按性别细分图 2.9，我们可以更好地理解数据。

```
fig = plt.figure(figsize=(18,6))
#Plotting gender based analysis for the survivals.
male = titanic_data.Survived[titanic_data.Sex == 'male'].value_counts().sort_index()
female = titanic_data.Survived[titanic_data.Sex == 'female'].value_counts().sort_index()
ax1 = fig.add_subplot(121)
male.plot(kind='barh',label='Male', alpha=0.55)
female.plot(kind='barh', color='#FA2379',label='Female', alpha=0.55)
plt.title("Gender analysis of survivals (raw value counts) ");
plt.legend(loc='best')
ax1.set_ylim(-1, 2)
ax2 = fig.add_subplot(122)
(male/float(male.sum())).plot(kind='barh',label='Male', alpha=0.55)
(female/float(female.sum())).plot(kind='barh',
color='#FA2379',label='Female', alpha=0.55)
plt.title("Gender analysis of survivals")plt.legend(loc='best')
ax2.set_ylim(-1, 2)
```

按类别细分后的数据如图 2.10 所示。

现在，我们有了关于两个可能类（生存和死亡）的更多信息。数据分析和可视化步骤是必要的，因为它可以让我们更深入地了解数据结构，并帮助我们为问题选择合适的学习算法。如你所见，我们从非常基本的图开始，然后增加图的复杂性以发现关于正在使用的数据的更多信息。

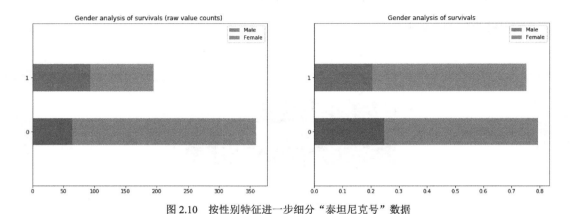

图 2.10　按性别特征进一步细分"泰坦尼克号"数据

2.3.2　数据分析——监督机器学习

本节旨在预测幸存者，因此，结果将是生存或死亡，这是一个二元分类问题：其中只有两个可能的类。

有很多学习算法可以用于二元分类问题，逻辑回归就是其中之一。正如维基百科所解释的：

在统计学中，逻辑回归是基于一个或多个预测变量预测可分类因变量的结果的一种回归分析方法（因变量可以取有限数目的值，其大小是没有意义的，但大小的排序可能是有意义的，也可能是没有意义的）。也就是说，逻辑回归用于估计定性输出模型中参数的实际值。描述单个试验的可能结果的概率可以建模为一个关于解释性（预测）变量的逻辑函数。"逻辑回归"常用于专门指代因变量是二元的问题，也就是说，可用类别的数量是两个，而两个以上类别的问题称为多项逻辑回归；如果类别是有序的，则称为有序逻辑回归。逻辑回归的目的是预测分类因变量与一个或多个自变量之间的关系，这些变量通常（但不一定）是连续的，但是可以使用概率值作为因变量的预测值。因此，解决这一类问题的时候可以使用和概率回归计算相似的方法。

为了使用逻辑回归，我们需要创建一个公式，以告诉模型特征/输入的类型。

```
# model formula
# here the ~ sign is an = sign, and the features of our dataset
```

2.3 "泰坦尼克号"示例——建立和训练模型

```
# are written as a formula to predict survived. The C() lets our
# regression know that those variables are categorical.
formula = 'Survived ~ C(Pclass) + C(Sex) + Age + SibSp + C(Embarked)'
# create a results dictionary to hold our regression results for easy
analysis later
results = {}
# create a regression friendly dataframe using patsy's dmatrices function
y,x = dmatrices(formula, data=titanic_data, return_type='dataframe')
# instantiate our model
model = sm.Logit(y,x)
# fit our model to the training data
res = model.fit()
# save the result for outputing predictions later
results['Logit'] = [res, formula]
res.summary()
Output:
Optimization terminated successfully.
         Current function value: 0.444388
         Iterations 6
```

输出结果如图 2.11 所示。

	Logit Regression Results			
Dep. Variable:	Survived	No. Observations:	712	
Model:	Logit	Df Residuals:	704	
Method:	MLE	Df Model:	7	
Date:	Sun, 20 Dec 2015	Pseudo R-squ.:	0.3414	
Time:	11:27:33	Log-Likelihood:	-316.40	
converged:	True	LL-Null:	-480.45	
		LLR p-value:	5.992e-67	

	coef	std err	z	P>|z|	[95.0% Conf. Int.]
Intercept	4.5423	0.474	9.583	0.000	3.613 5.471
C(Pclass)[T.2]	-1.2673	0.299	-4.245	0.000	-1.852 -0.682
C(Pclass)[T.3]	-2.4966	0.296	-8.422	0.000	-3.078 -1.916
C(Sex)[T.male]	-2.6239	0.218	-12.060	0.000	-3.050 -2.197
C(Embarked)[T.Q]	-0.8351	0.597	-1.398	0.162	-2.006 0.335
C(Embarked)[T.S]	-0.4254	0.271	-1.572	0.116	-0.956 0.105
Age	-0.0436	0.008	-5.264	0.000	-0.060 -0.027
SibSp	-0.3697	0.123	-3.004	0.003	-0.611 -0.129

图 2.11 逻辑回归的结果

下面绘制模型与实际模型的预测值与残差（即目标变量的实际值与预测值之间的差值）。

```
# Plot Predictions Vs Actual
plt.figure(figsize=(18,4))
plt.subplot(121, axisbg="#DBDBDB")
# generate predictions from our fitted model
ypred = res.predict(x)
plt.plot(x.index, ypred, 'bo', x.index, y, 'mo', alpha=.25)
plt.grid(color='white', linestyle='dashed')
plt.title('Logit predictions, Blue: \nFitted/predicted values: Red')
# Residuals
ax2 = plt.subplot(122, axisbg="#DBDBDB")
plt.plot(res.resid_dev, 'r-')
plt.grid(color='white', linestyle='dashed')
ax2.set_xlim(-1, len(res.resid_dev))
plt.title('Logit Residuals')
```

绘制结果如图 2.12 所示。

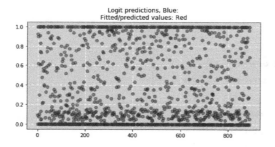 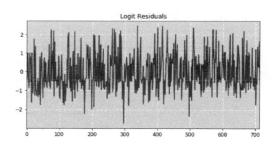

图 2.12　更深入地了解逻辑回归模型

现在，我们已经构建了逻辑回归模型，在此之前，我们已经对数据集进行了一些分析和探讨。前面的例子显示了构建机器学习解决方案的一般流程。

很多时候，从业者会陷入一些技术陷阱，因为他们缺乏对机器学习概念的理解。例如，有人可能在测试集中获得 99% 的准确性，然后在没有对数据中类的分布进行调查（例如有多少样本是负数以及有多少样本是正数）的情况下就部署了模型。

为了突出其中的一些概念，并区分我们需要注意的以及那些应该真正关心的误差，下面将介绍误差分析。

2.4 不同类型的误差解析

在机器学习中，有两种类型的误差，作为数据科学的新手，我们需要了解它们之间的关键差异。如果最终将误差类型的损失函数最小化，那么整个学习系统将是无用的，我们将无法在实践中在未见过的数据上使用它。为了尽量减少从业者对这两类误差的这种误解，以下两节将对它们进行解释。

2.5 表现（训练集）误差

表现误差是我们面临的第一种类型的误差。通过训练得到此类误差的较小值并不意味着读者的模型将在未知数据上会更好地分类（泛化）。为了更好地理解这种类型的误差，这里将给出一个有关分类问题的简单例子。在课堂上解决问题的目的不是要在考试中再次解决同样的问题，而是为了能够解决其他问题，这些问题不一定与我们在课堂上解决的问题类似。考试题目可能来自同一系列的课堂练习，但不一定完全相同。

表现误差是训练模型在真实结果/输出已知的训练集上表现的能力。如果设法在训练集上实现零错误，那么对读者来说，这表明模型（大部分情况下）对于之前没见过的数据将不能很好地分类（不会泛化），另一方面，在数据科学中将训练集作为学习算法的基础知识，目的是很好地处理未来看不见的数据。

在图2.13中，上方曲线表示**表现**误差。每次增加模型记忆事物的能力（例如，通过增加解释性特征的数量来增加模型复杂性）时，我们都会发现这个表现误差将接近于零。可以证明，如果你有与样本一样多的特征，那么**表现**误差将为零。

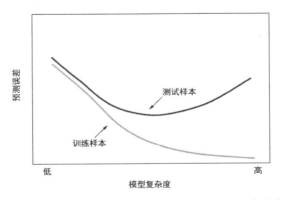

图2.13 表现误差（粗黑曲线）和泛化/真实误差（浅灰色曲线）

2.6 泛化/真实误差

泛化/真实误差是数据科学中的第二种误差类型,也是更重要的误差类型。构建学习系统的整体目的是能够在测试集上获得较小的泛化误差。换句话说,使模型在一组尚未在训练阶段使用的样本中很好地进行判断。如果我们仍然考虑 2.5 节中的课程场景,那么可以将泛化误差视为解决考试问题的能力,这些问题不一定与我们在校学习的学科中解决的问题类似。因此,泛化性能是为了正确预测未知数据的结果/输出模型使用在训练阶段学到的技能(参数)的能力。

在图 2.13 中,浅灰色曲线表示泛化误差。可以看到,随着模型复杂性的增加,泛化误差将会减小,直到模型失去降低泛化误差的能力,并且泛化误差不再减小。导致泛化误差的泛化能力不再增加的这一部分曲线,称为**过拟合曲线**。

本节的主旨是尽可能地减小泛化误差。

2.7 总结

线性模型是一个非常强大的工具,如果我们的数据与其假设相匹配,可以将它用作初始学习算法。理解线性模型将帮助读者理解将线性模型作为基础构件的更为复杂的模型。

接下来,我们将继续使用"泰坦尼克号"示例更详细地解析模型的复杂性以及如何评估模型。模型复杂性是一个非常强大的工具,读者需要谨慎地操作它以增强模型的泛化能力,误用它会导致过度拟合问题。

第 3 章
特征工程与模型复杂性——重温"泰坦尼克号"示例

模型复杂性与评估是构建一个成功的数据科学系统必不可少的一步。有很多工具可以用来评估和选择模型。本章将讨论一些可以帮助读者通过添加更多描述性特征并从现有数据中提取有用信息来提高数据价值的工具。同样地,本章也会讨论与最优数量特征有关的其他工具,并了解为什么"有大量的特征却只有很少量的训练样本/观测值"会是一个问题。

以下是本章主要阐述的主题。

- 特征工程。
- 维度灾难。
- 重温"泰坦尼克号"示例——融会贯通。
- 偏差-方差分解。
- 学习可见性。

3.1 特征工程

特征工程是有助于提高模型性能的关键组件之一。一个选择了正确特征的简单模型往往比那些选择了较差特征的复杂模型表现得更好。读者可以将特征工程视为决定预测模型成败与否的最重要的一步。如果对数据比较了解,特征工程将会变得更加容易。

第 3 章 特征工程与模型复杂性——重温"泰坦尼克号"示例

任何使用机器学习来解决某个特定问题的人都会广泛使用特征工程,这个问题就是:**如何充分地利用数据样本来进行模型预测?** 这是特征工程的过程和实践所要解决的问题,数据科学技能的成功往往始于知道如何更好地表示数据。

预测建模是将一系列特征或者输入变量$(x_1, x_2, \cdots, x_n)$转换成感兴趣的输出或者目标(y)的一种方案或规则。那么,什么是特征工程呢?它是从现有输入变量$(x_1, x_2, \cdots, x_n)$创建新的输入变量或特征$(z_1, z_2, \cdots, z_n)$的过程。这里不是随便创建一些新的特征,新建的特征应该对模型的输出有贡献并且与之相关。有了相关领域(如市场营销、医疗等领域)的知识后,创建与模型输出有关的这些特征将是一个很容易的过程。如果机器学习从业者在此过程中与一些领域专家相互交流,特征工程的结果将会变得更好。

相关领域知识可能对特征工程有很大帮助的一个例子就是,给定一组输入变量/特征(如温度、风速和云量百分比)来分析下雨的可能性。对于这个具体的例子,可以构建一个新的名为 overcast 的二元特征。当云量百分比小于 20% 时,overcast 的值为 1 或者 no;否则,overcast 的值等于 0 或者 yes。在这个例子中,相关专业知识对于指定阈值或截断百分比是十分重要的,选择的输入越合理、越有用,最终模型的可靠性越高,预测性越准确。

3.1.1 特征工程的类型

作为一种技术,特征工程有 3 种重要的子类别。作为一名深度学习实践者,读者完全可以自由选择其中一种或者以某种方式对它们进行组合。

1. 特征选择

特征重要性是根据输入变量对目标/输出变量的贡献度将它们排序的过程。另外,可以认为这个过程是根据输入变量在模型预测能力中的价值来对它们进行排序的过程。

一些学习方法(如决策树)将这种特征排名或重要性作为其内部过程的一部分。大多数情况下,这些方法使用熵来筛掉一些不太有价值的变量。在某些情况下,深度学习实践者使用这些学习方法来选择最重要的特征,然后将它们输入更好的学习算法中。

2. 降维

降维有时也可以称为特征提取,它是一种将现有输入变量组合为一组数目大量减少的新输入变量的过程。这类特征工程中一种最常用的方法就是**主成分分析**(Principle Component Analysis,PCA),该算法利用数据的方差来生成一组和原始输入变量不太相同的、数目减少的新输入变量。

3. 特征构造

特征构造是一种常用的特征工程类型，大家在讨论特征工程时经常会提到它，该方法是手动处理原始数据或构建新特征的过程。在这种类型的特征工程中，领域知识对于从现有特征构建其他新特征是非常有帮助的。像其他特征工程技术一样，特征构造的目的也是增加模型的预测能力。特征构造的一个简单例子就是使用时间戳来产生两个新特征，如 AM 和 PM，这对于区分白天和夜晚很有帮助。同样，也可以通过计算噪声特征的均值然后确定给定数据行是大于还是小于该均值，从而将噪声数值特征转换为更简单的名义特征。

3.1.2 重温"泰坦尼克号"示例

在本节中，读者将再次看到"泰坦尼克"示例，但是这里会使用特征工程的工具从另一个视角解读。若读者跳过了第 2 章，那么这里再说明一下，"泰坦尼克号"示例是 Kaggle 上面的一个竞赛，其目的是预测特定乘客是否[1]生还。

在重新看这个例子的时候，我们将会使用到 scikit-learn 和 pandas 库。首先，看训练集和测试集，获取一些关于数据的统计信息。

```
# reading the train and test sets using pandas
train_data = pd.read_csv('data/train.csv', header=0)
test_data = pd.read_csv('data/test.csv', header=0)

# concatenate the train and test set together for doing the overall feature
engineering stuff
df_titanic_data = pd.concat([train_data, test_data])

# removing duplicate indices due to coming the train and test set by reindexing
the data
df_titanic_data.reset_index(inplace=True)

# removing the index column the reset_index() function generates
df_titanic_data.drop('index', axis=1, inplace=True)

# index the columns to be 1-based index
df_titanic_data = df_titanic_data.reindex_axis(train_data.columns, axis=1)
```

---

[1] 原文中此处错误地将 whether（是否）写成了 weather（天气）。——译者注

这里需要指出一些关于以上代码片段的知识点。

- 以上代码使用了 pandas 库的 concat 函数来联合训练集和测试集的数据框架。这对于特征工程任务很有帮助,因为需要全面了解输入变量/特征的分布。
- 在结合了两个数据框架之后,还需要对输出的数据框架做一些修改。

1. 缺失值处理

从客户那里获取新数据集后,缺失值处理将是首先要考虑的事情,因为几乎每个数据集都会存在缺失或不正确的数据。在接下来的章节中,我们可以看到一些学习算法能够处理缺失值问题,但其他学习算法需要读者自己处理缺失数据。这个例子使用的是 scikit-learn 库中的随机森林分类器,这个算法需要单独处理缺失数据。

读者可以使用各种不同的方法来处理缺失数据。

(1) 删除所有存在缺失值的样本

如果现有数据集比较小而且存在很多缺失值,则删除所有存在缺失值的样本不是一个好的选择,因为删除这些存在缺失值的样本将会使得数据没有意义。但是如果现有数据集很大,删除存在缺失值样本对原始数据集没有多大影响,那它会是一个快速而简单的方法。

(2) 缺失值输入

当数据是分类数据时,输入缺失值很有用。这种方法背后的逻辑是缺失值可能和其他变量相关联,删除它们可能会导致信息丢失,这会影响到模型的性能。

例如,对于一个可能取值为-1 或 1 的二元变量,可以考虑给它添加另一个值(0)来表示缺失值。可以用 U0 来替代 Cabin 特征中的空值,代码如下所示。

```
# replacing the missing value in cabin variable "U0"
df_titanic_data['Cabin'][df_titanic_data.Cabin.isnull()] = 'U0'
```

(3) 分配均值

因为分配均值比较简单,所以它也是一种常用的方法。对于数值特征,可以使用均值或者中位数来代替缺失值。同样也可以在类别变量的情况下使用这种方法,具体操作是用众数(出现频率最高的值)来替换缺失值。

下面这段代码将 Fare 特征中非缺失值的中位数赋值给缺失值。

```
# handling the missing values by replacing it with the median fare
df_titanic_data['Fare'][np.isnan(df_titanic_data['Fare'])] =
df_titanic_data['Fare'].median()
```

或者可以使用以下代码来求 Embarked 特征中出现次数最多的值，并将它赋值给缺失值。

```
# replacing the missing values with the most common value in the variable
df_titanic_data.Embarked[df_titanic_data.Embarked.isnull()] =
df_titanic_data.Embarked.dropna().mode().values
```

（4）使用回归或者其他简单模型预测缺失变量的值

在处理"泰坦尼克号"示例中的 Age 特征时用到了这种方法。Age 特征是预测乘客是否生还很重要的一点，如果采用前面提到的通过均值来处理的方法将会丢失一些信息。

为了预测缺失值，这里需要用到监督学习算法，该算法以可用特征作为输入，以想要预测缺失值的那些特征的可能值作为输出。下面的代码使用随机森林分类器来预测 Age 特征中的缺失值。

```
# Define a helper function that can use RandomForestClassifier for handling
the missing values of the age variable
def set_missing_ages():
    global df_titanic_data

    age_data = df_titanic_data[
        ['Age', 'Embarked', 'Fare', 'Parch', 'SibSp', 'Title_id', 'Pclass',
'Names', 'CabinLetter']]
    input_values_RF =
age_data.loc[(df_titanic_data.Age.notnull())].values[:, 1::]
    target_values_RF =
age_data.loc[(df_titanic_data.Age.notnull())].values[:, 0]

    # Creating an object from the random forest regression function of
sklearn<use the documentation for more details>
    regressor = RandomForestRegressor(n_estimators=2000, n_jobs=-1)

    # building the model based on the input values and target values above
    regressor.fit(input_values_RF, target_values_RF)

    # using the trained model to predict the missing values
```

```
    predicted_ages = 
regressor.predict(age_data.loc[(df_titanic_data.Age.isnull())].values[:,1::])
    # Filling the predicted ages in the original titanic dataframe
    age_data.loc[(age_data.Age.isnull()), 'Age'] = predicted_ages
```

2. 特征转换

前两节介绍了读取训练集和测试集并将它们合并起来的相关知识，同时也处理了一些缺失值。现在，我们可以使用 scikit-learn 库中的随机森林分类器来预测乘客的生存率。随机森林算法的不同实现方法可能接受不同类型的数据。scikit-learn 库中的随机森林算法的实现只接受数字数据，所以这里需要将类别特征转换为数字特征。

特征有以下两种类型。

- **定量特征**：用数值尺度来度量，它能够进行有意义的排序。在 Titanic 数据样本中，Age 特征就是定量特征。
- **定性变量**：也叫作**分类**（categorical）**变量**，这些变量不是数字形式的，它们描述符合类别的数据。在 Titanic 数据样本中，Embarked（表明出发港名称）特征就是定性特征。

我们可以对不同的变量采用不同的转换方法。下面是一些用来转换量化/类别特征的方法。

（1）虚拟特征

这些变量也称为类别特征或者二元特征。如果需要转换的特征只有少量不同的取值，虚拟特征会是一个不错的选择。在 Titanic 数据样本中，Embarked 特征只有 3 种不同值（S、C 和 Q），它们频繁出现。所以，我们可以将 Embarked 特征转换为 3 个虚拟变量（Embarked_S、Embarked_C 和 Embarked_Q），以便能够使用随机森林分类器。

下面的代码将展示如何进行这种转化。

```
# constructing binary features
def process_embarked():
    global df_titanic_data
    
    # replacing the missing values with the most common value in the variable
    df_titanic_data.Embarked[df.Embarked.isnull()] = 
df_titanic_data.Embarked.dropna().mode().values
```

```
    # converting the values into numbers
    df_titanic_data['Embarked'] =
pd.factorize(df_titanic_data['Embarked'])[0]

    # binarizing the constructed features
    if keep_binary:
        df_titanic_data = pd.concat([df_titanic_data,
pd.get_dummies(df_titanic_data['Embarked']).rename(
            columns=lambda x: 'Embarked_' + str(x))], axis=1)
```

（2）因式分解

因式分解用于从任何其他特征中产生数字类别特征。在 pandas 库中，factorized()函数可以做到这一点。如果特征是字母数值类别特征，则这种类型的转换就很有用。在 Titanic 数据样本中，我们可以将 Cabin 特征转换为类别特征，代表舱室的字母。

```
# the cabin number is a sequence of of alphanumerical digits, so we are
going to create some features
# from the alphabetical part of it
df_titanic_data['CabinLetter'] = df_titanic_data['Cabin'].map(lambda l:
get_cabin_letter(l))
df_titanic_data['CabinLetter'] =
pd.factorize(df_titanic_data['CabinLetter'])[0]

def get_cabin_letter(cabin_value):
    # searching for the letters in the cabin alphanumerical value
    letter_match = re.compile("([a-zA-Z]+)").search(cabin_value)

    if letter_match:
        return letter_match.group()
    else:
        return 'U'
```

读者还可以通过使用以下方法将变换应用于定量特征。

（3）缩放

缩放转换只能应用于数字特征。

例如，在 Titanic 数据样本中，Age 特征可以达到 100，但是家庭收入可能数百万美元计。有些模型对数值的大小很敏感，所以缩放这些特征有助于这些模型取得更好的表现。此外，缩放可以将变量的值压缩到特定范围内。

下面的代码通过对每个值减去样本均值来缩放 Age 特征，然后将它们缩放到单位方差。

```
# scale by subtracting the mean from each value
scaler_processing = preprocessing.StandardScaler()
df_titanic_data['Age_scaled'] =
scaler_processing.fit_transform(df_titanic_data['Age'])
```

（4）分档

分档这种定量转换用于创建分位数。在这种情况下，把定量特征值变换为有序变量。这种方法对于线性回归并不是一个好的选择，但是对于当使用有序变量或类别变量时能够有效响应的学习算法来说，这种方法可能会比较有效。

下面的代码将这种变换方法应用于 Fare 特征。

```
# Binarizing the features by binning them into quantiles
df_titanic_data['Fare_bin'] = pd.qcut(df_titanic_data['Fare'], 4)

if keep_binary:
    df_titanic_data = pd.concat(
        [df_titanic_data,
pd.get_dummies(df_titanic_data['Fare_bin']).rename(columns=lambda x:
'Fare_' + str(x))],
        axis=1)
```

3. 派生特征

在前一节中，我们对 Titanic 数据进行了一些转换，以便能够使用 scikit-learn 库（只接受数字数据）中的随机森林分类器。在本节中，我们将会定义另一种类型的变量，它派生自一个或多个其他特征。

在这个定义下，我们可以说之前的一些变换也称为**派生特征**。本节将会讨论其他复杂的变换。

前面提到过读者需要使用自己的特征工程技能来获取新的特征以增强模型的预测能力，同样也谈到了特征工程在数据科学流程中的重要性，以及为什么需要花费大部分的时间和精力来考虑一些有用的特征。在本节中领域知识是非常有用的。

派生特征的一个非常简单的例子，类似于从电话号码中提取出国家代码或者区域代码，或者从 GPS 坐标中提取国家信息或区域信息。

Titanic 数据是非常简单的数据，并没有包含很多可用的变量，但是读者可以尝试从现有的文本特征中派生出一些其他特征。

（1）Name（名称）

name 变量对于大多数数据集来说都是无用的，但是它有两个有用的特性。第一个是名字的长度。例如，名字的长度可能会反映出你的身份，因此也会影响到你登上救生艇的可能性。

```
# getting the different names in the names variable
df_titanic_data['Names'] = df_titanic_data['Name'].map(lambda y:
len(re.split(' ', y)))
```

第二个有趣的属性就是名字的 Title，它能够用来判断你的身份或者性别信息。

```
# Getting titles for each person
df_titanic_data['Title'] = df_titanic_data['Name'].map(lambda y:
re.compile(", (.*?)\.").findall(y)[0])

# handling the low occurring titles
df_titanic_data['Title'][df_titanic_data.Title == 'Jonkheer'] = 'Master'
df_titanic_data['Title'][df_titanic_data.Title.isin(['Ms', 'Mlle'])] =
'Miss'
df_titanic_data['Title'][df_titanic_data.Title == 'Mme'] = 'Mrs'
df_titanic_data['Title'][df_titanic_data.Title.isin(['Capt', 'Don',
'Major', 'Col', 'Sir'])] = 'Sir'
df_titanic_data['Title'][df_titanic_data.Title.isin(['Dona', 'Lady', 'the
Countess'])] = 'Lady'

# binarizing all the features
if keep_binary:
    df_titanic_data = pd.concat(
        [df_titanic_data,
pd.get_dummies(df_titanic_data['Title']).rename(columns=lambda x: 'Title_'
+ str(x))],
        axis=1)
```

我们还可以尝试从 Name 这个特征中提出一些其他的有趣的特征。例如，有些人可能会想到通过名字中的姓氏特征来查明"泰坦尼克号"游轮上该姓氏的家庭成员的个数等。

(2) Cabin(船舱)

在 Titanic 数据中,Cabin 特征由一个表示甲板的字母和一个表示房间号的数字组成。房间号码从船头向船尾递增,这非常有助于了解乘客位置信息。同样,从不同的甲板信息也可以获取乘客的状态,这也有助于确定哪些人登上了救生艇。

```
# repllacing the missing value in cabin variable "U0"
df_titanic_data['Cabin'][df_titanic_data.Cabin.isnull()] = 'U0'

# the cabin number is a sequence of of alphanumerical digits, so we are
going to create some features
# from the alphabetical part of it
df_titanic_data['CabinLetter'] = df_titanic_data['Cabin'].map(lambda l:
get_cabin_letter(l))
df_titanic_data['CabinLetter'] =
pd.factorize(df_titanic_data['CabinLetter'])[0]

# binarizing the cabin letters features
if keep_binary:
    cletters =
pd.get_dummies(df_titanic_data['CabinLetter']).rename(columns=lambda x:
'CabinLetter_' + str(x))
    df_titanic_data = pd.concat([df_titanic_data, cletters], axis=1)

# creating features from the numerical side of the cabin
df_titanic_data['CabinNumber'] = df_titanic_data['Cabin'].map(lambda x:
get_cabin_num(x)).astype(int) + 1
```

(3) Ticket(票)

Ticket 特征的代码不是非常清晰,但我们可以做一些猜测并尝试对它分组。在查看了 Ticket 特征后,可以得到如下线索。

- 近 1/4 的船票都是以字符开头的,而剩下的 3/4 都只包含数字。
- 票号的数字部分似乎隐含着一些关于乘客类别的信息。例如,以 "1" 开头的船票通常是头等票,以 "2" 开头的船票通常是二等票,以 "3" 开头的船票通常是三等票。这里只说大部分情况下,因为它适用于大部分情况,但并不是全部。还有一些票号是以数字 4~9 开头的,但这些票很少,几乎只有三等票。
- 有些人共享一个票号,这可能表明他们来自一个家庭或者是一起旅行并且像一家人的关系亲密的朋友。

以下这段代码尝试分析船票的特征码来提供一些新的线索。

```
# Helper function for constructing features from the ticket variable
def process_ticket():
    global df_titanic_data

    df_titanic_data['TicketPrefix'] = df_titanic_data['Ticket'].map(lambda y: get_ticket_prefix(y.upper()))
    df_titanic_data['TicketPrefix'] = df_titanic_data['TicketPrefix'].map(lambda y: re.sub('[\.?\/?]', '', y))
    df_titanic_data['TicketPrefix'] = df_titanic_data['TicketPrefix'].map(lambda y: re.sub('STON', 'SOTON', y))

    df_titanic_data['TicketPrefixId'] = pd.factorize(df_titanic_data['TicketPrefix'])[0]

    # binarzing features for each ticket layer
    if keep_binary:
        prefixes = pd.get_dummies(df_titanic_data['TicketPrefix']).rename(columns=lambda y: 'TicketPrefix_' + str(y))
        df_titanic_data = pd.concat([df_titanic_data, prefixes], axis=1)

    df_titanic_data.drop(['TicketPrefix'], axis=1, inplace=True)

    df_titanic_data['TicketNumber'] = df_titanic_data['Ticket'].map(lambda y: get_ticket_num(y))
    df_titanic_data['TicketNumberDigits'] = df_titanic_data['TicketNumber'].map(lambda y: len(y)).astype(np.int)
    df_titanic_data['TicketNumberStart'] = df_titanic_data['TicketNumber'].map(lambda y: y[0:1]).astype(np.int)
    df_titanic_data['TicketNumber'] = df_titanic_data.TicketNumber.astype(np.int)

    if keep_scaled:
        scaler_processing = preprocessing.StandardScaler()
        df_titanic_data['TicketNumber_scaled'] = scaler_processing.fit_transform(
            df_titanic_data.TicketNumber.reshape(-1, 1))

def get_ticket_prefix(ticket_value):
```

```python
    # searching for the letters in the ticket alphanumerical value
    match_letter = re.compile("([a-zA-Z\.\/]+)").search(ticket_value)
    if match_letter:
        return match_letter.group()
    else:
        return 'U'

def get_ticket_num(ticket_value):
    # searching for the numbers in the ticket alphanumerical value
    match_number = re.compile("([\d]+$)").search(ticket_value)
    if match_number:
        return match_number.group()
    else:
        return '0'
```

4. 交互特征

通过对特征集合执行一些数学运算并指出变量之间关系的作用，我们可以获得一些交互特征。这里在数字特征上使用基本的数学运算，来查看变量之间关系的作用。

```
# Constructing features manually based on the interaction between the
individual features
numeric_features = df_titanic_data.loc[:,
                    ['Age_scaled', 'Fare_scaled', 'Pclass_scaled',
'Parch_scaled', 'SibSp_scaled',
                     'Names_scaled', 'CabinNumber_scaled',
'Age_bin_id_scaled', 'Fare_bin_id_scaled']]
print("\nUsing only numeric features for automated feature generation:\n",
numeric_features.head(10))

new_fields_count = 0
for i in range(0, numeric_features.columns.size - 1):
    for j in range(0, numeric_features.columns.size - 1):
        if i <= j:
            name = str(numeric_features.columns.values[i]) + "*" +
str(numeric_features.columns.values[j])
            df_titanic_data = pd.concat(
                [df_titanic_data, pd.Series(numeric_features.iloc[:, i] *
numeric_features.iloc[:, j], name=name)],
                axis=1)
            new_fields_count += 1
        if i < j:
```

```
                name = str(numeric_features.columns.values[i]) + "+" +
str(numeric_features.columns.values[j])
                df_titanic_data = pd.concat(
                    [df_titanic_data, pd.Series(numeric_features.iloc[:, i] +
numeric_features.iloc[:, j], name=name)],
                    axis=1)
                new_fields_count += 1
            if not i == j:
                name = str(numeric_features.columns.values[i]) + "/" +
str(numeric_features.columns.values[j])
                df_titanic_data = pd.concat(
                    [df_titanic_data, pd.Series(numeric_features.iloc[:, i] /
numeric_features.iloc[:, j], name=name)],
                    axis=1)
                name = str(numeric_features.columns.values[i]) + "-" +
str(numeric_features.columns.values[j])
                df_titanic_data = pd.concat(
                    [df_titanic_data, pd.Series(numeric_features.iloc[:, i] -
numeric_features.iloc[:, j], name=name)],
                    axis=1)
                new_fields_count += 2

print("\n", new_fields_count, "new features constructed")
```

这种特征工程可以产生很多特征。之前的代码片段使用了 9 种基本特征来产生 176 种交互特征。

也可以考虑删除一些高度相关的特征，因为这些特征的存在并不会给模型增加任何其他的信息。这里可以使用 Spearman 相关性来识别和删除高度相关的特征。Spearman 方法的输出包含一个等级系数，该系数可以用来识别高度相关的特征。

```
# using Spearman correlation method to remove the feature that have high
Correlation

# calculating the correlation matrix
df_titanic_data_cor = df_titanic_data.drop(['Survived', 'PassengerId'],
axis=1).corr(method='spearman')

# creating a mask that will ignore correlated ones
mask_ignore = np.ones(df_titanic_data_cor.columns.size) -
np.eye(df_titanic_data_cor.columns.size)
```

```
df_titanic_data_cor = mask_ignore * df_titanic_data_cor

features_to_drop = []

# dropping the correclated features
for column in df_titanic_data_cor.columns.values:

    # check if we already decided to drop this variable
    if np.in1d([column], features_to_drop):
        continue

    # finding highly correlacted variables
    corr_vars = df_titanic_data_cor[abs(df_titanic_data_cor[column])>
0.98].index
    features_to_drop = np.union1d(features_to_drop, corr_vars)

print("\nWe are going to drop", features_to_drop.shape[0], " which are
highly correlated features...\n")
df_titanic_data.drop(features_to_drop, axis=1, inplace=True)
```

3.2 维度灾难

为了更好地解释维度灾难和过拟合的问题，这里将通过一个有一组图像的例子来进行说明。每一幅图像中都有一只猫或一只狗。于是，我们希望建立一个模型以区分包含猫和包含狗的图像。就像第 1 章中的鱼类识别系统一样，我们需要找到一个解释性特征，学习算法能够使用它来区分两类动物（猫和狗）。在这个例子中，色彩是一个区分猫和狗的良好描述符。我们可以将红色的平均值、蓝色的平均值和绿色的平均值作为解释性特征来区分这两类动物。

该算法接下来会以某种方式结合这 3 种特征，以形成这两个类之间的决策边界。

这 3 种特征的一个简单线性组合可以表示成如下形式。

```
If 0.5*red + 0.3*green + 0.2*blue > 0.6 : return cat;

else return dog;
```

因为这些描述性特征还不足以获得一个好的分类，所以我们可以考虑增加更多能够增强模型区分猫和狗的预测能力的特征。例如，可以考虑通过计算图像在 X 和 Y 维度的平均边缘与梯度来增加诸如图像纹理等特征。在增加这两个特征后，模型的准确性会提高。我们还可以通过添加越来越多基于颜色、纹理直方图、统计时刻等的特征来提高模型（分类器）分类能力。读者可以很轻松地添加几百个这样的特征来增强模型的预测性能，但是当增加的特征超出一定限制之后，结果反而会变得更糟。通过图 3.1，可以更好地理解这一点。

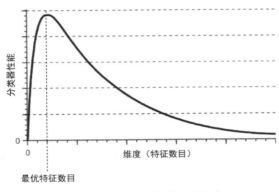

图 3.1 模型性能与特征数目的关系

如图 3.1 所示，随着特征数目的增加，分类器的性能也随之提高，直到达到最佳性能。但在同样大小的训练集上，如果再继续增加特征，分类器的性能将会变差。

避免维度灾难

在前面的内容中，我们可以看到，当特征数量超过某个最优值之后，分类器的性能将会下降。从理论上讲，如果拥有无限的训练样本，那么维度灾难是不存在的。所以，最优特征数量完全取决于数据集的大小。

一种有助于避免这种维度灾难的方法是从 N 个特征中抽取出含有 M 个特征的子集，其中 $M \ll N$。M 中的每个特征都可以是 N 中某些特征的组合。有一些现成算法可以做到这一点，这些算法以某种方式试图找到原始的 N 个特征中有用的、不相关的、线性的组合。其中常用的技术就是**主成分分析**（Principle Component Analysis，PCA）。PCA 试图找到最少数量的特征，从而能够捕获原始数据的最大方差。

将 PCA 应用于原始训练特征的有用且简单的代码如下。

```python
# minimum variance percentage that should be covered by the reduced number
of variables
variance_percentage = .99

# creating PCA object
pca_object = PCA(n_components=variance_percentage)

# trasforming the features
input_values_transformed = pca_object.fit_transform(input_values,
target_values)

# creating a datafram for the transformed variables from PCA
pca_df = pd.DataFrame(input_values_transformed)

print(pca_df.shape[1], " reduced components which describe ",
str(variance_percentage)[1:], "% of the variance")
```

在"泰坦尼克号"示例中，我们试图在对原始特征应用和不应用 PCA 的情况下建立分类器。因为最终采用的是随机森林分类器，所以应用 PCA 对结果很有帮助。随机森林在没有任何特征变换的情况下表现良好，甚至相关的特征并没有对模型产生多少影响。

3.3 重温"泰坦尼克号"示例——融会贯通

本节将会把特征工程和降维的相关知识与技巧放在一起讲解。该示例的全部代码如下。

```python
import re
import numpy as np
import pandas as pd
import random as rd
from sklearn import preprocessing
from sklearn.cluster import KMeans
from sklearn.ensemble import RandomForestRegressor
from sklearn.decomposition import PCA

# Print options
np.set_printoptions(precision=4, threshold=10000, linewidth=160,
edgeitems=999, suppress=True)
pd.set_option('display.max_columns', None)
pd.set_option('display.max_rows', None)
```

```python
pd.set_option('display.width', 160)
pd.set_option('expand_frame_repr', False)
pd.set_option('precision', 4)


# constructing binary features
def process_embarked():
    global df_titanic_data

    # replacing the missing values with the most common value in the variable
    df_titanic_data.Embarked[df.Embarked.isnull()] = df_titanic_data.Embarked.dropna().mode().values

    # converting the values into numbers
    df_titanic_data['Embarked'] = pd.factorize(df_titanic_data['Embarked'])[0]

    # binarizing the constructed features
    if keep_binary:
        df_titanic_data = pd.concat([df_titanic_data, pd.get_dummies(df_titanic_data['Embarked']).rename(
            columns=lambda x: 'Embarked_' + str(x))], axis=1)

# Define a helper function that can use RandomForestClassifier for handling
# the missing values of the age variable
def set_missing_ages():
    global df_titanic_data

    age_data = df_titanic_data[
        ['Age', 'Embarked', 'Fare', 'Parch', 'SibSp', 'Title_id', 'Pclass', 'Names', 'CabinLetter']]
    input_values_RF = age_data.loc[(df_titanic_data.Age.notnull())].values[:, 1::]
    target_values_RF = age_data.loc[(df_titanic_data.Age.notnull())].values[:, 0]

    # Creating an object from the random forest regression function of sklearn<use the documentation for more details>
    regressor = RandomForestRegressor(n_estimators=2000, n_jobs=-1)

    # building the model based on the input values and target values above
    regressor.fit(input_values_RF, target_values_RF)
```

```python
    # using the trained model to predict the missing values
    predicted_ages = 
regressor.predict(age_data.loc[(df_titanic_data.Age.isnull())].values[:,
1::])

    # Filling the predicted ages in the original titanic dataframe
    age_data.loc[(age_data.Age.isnull()), 'Age'] = predicted_ages

# Helper function for constructing features from the age variable
def process_age():
    global df_titanic_data

    # calling the set_missing_ages helper function to use random forest
regression for predicting missing values of age
    set_missing_ages()

    # # scale the age variable by centering it around the mean with a unit
variance
    #     if keep_scaled:
    #         scaler_preprocessing = preprocessing.StandardScaler()
    #         df_titanic_data['Age_scaled'] =
scaler_preprocessing.fit_transform(df_titanic_data.Age.reshape(-1, 1))

    # construct a feature for children
    df_titanic_data['isChild'] = np.where(df_titanic_data.Age < 13, 1, 0)

    # bin into quartiles and create binary features
    df_titanic_data['Age_bin'] = pd.qcut(df_titanic_data['Age'], 4)
    if keep_binary:
        df_titanic_data = pd.concat([df_titanic_data,
pd.get_dummies(df_titanic_data['Age_bin']).rename(columns=lambda y: 'Age_'
+ str(y))], axis=1)

    if keep_bins:
        df_titanic_data['Age_bin_id'] =
pd.factorize(df_titanic_data['Age_bin'])[0] + 1

    if keep_bins and keep_scaled:
        scaler_processing = preprocessing.StandardScaler()
        df_titanic_data['Age_bin_id_scaled'] =
scaler_processing.fit_transform(
            df_titanic_data.Age_bin_id.reshape(-1, 1))
```

```python
    if not keep_strings:
        df_titanic_data.drop('Age_bin', axis=1, inplace=True)

# Helper function for constructing features from the passengers/crew names
def process_name():
    global df_titanic_data

    # getting the different names in the names variable
    df_titanic_data['Names'] = df_titanic_data['Name'].map(lambda y: len(re.split(' ', y)))

    # Getting titles for each person
    df_titanic_data['Title'] = df_titanic_data['Name'].map(lambda y: re.compile(", (.*?)\.").findall(y)[0])

    # handling the low occurring titles
    df_titanic_data['Title'][df_titanic_data.Title == 'Jonkheer'] = 'Master'
    df_titanic_data['Title'][df_titanic_data.Title.isin(['Ms', 'Mlle'])] = 'Miss'
    df_titanic_data['Title'][df_titanic_data.Title == 'Mme'] = 'Mrs'
    df_titanic_data['Title'][df_titanic_data.Title.isin(['Capt', 'Don', 'Major', 'Col', 'Sir'])] = 'Sir'
    df_titanic_data['Title'][df_titanic_data.Title.isin(['Dona', 'Lady', 'the Countess'])] = 'Lady'

    # binarizing all the features
    if keep_binary:
        df_titanic_data = pd.concat([df_titanic_data, pd.get_dummies(df_titanic_data['Title']).rename(columns=lambda x: 'Title_' + str(x))], axis=1)

    # scaling
    if keep_scaled:
        scaler_preprocessing = preprocessing.StandardScaler()
        df_titanic_data['Names_scaled'] = scaler_preprocessing.fit_transform(df_titanic_data.Names.reshape(-1, 1))

    # binning
    if keep_bins:
        df_titanic_data['Title_id'] = pd.factorize(df_titanic_data['Title'])[0] + 1
```

```python
    if keep_bins and keep_scaled:
        scaler = preprocessing.StandardScaler()
        df_titanic_data['Title_id_scaled'] =
scaler.fit_transform(df_titanic_data.Title_id.reshape(-1, 1))

# Generate features from the cabin input variable
def process_cabin():
    # refering to the global variable that contains the titanic examples
    global df_titanic_data

    # repllacing the missing value in cabin variable "U0"
    df_titanic_data['Cabin'][df_titanic_data.Cabin.isnull()] = 'U0'

    # the cabin number is a sequence of of alphanumerical digits, so we are going to create some features
    # from the alphabetical part of it
    df_titanic_data['CabinLetter'] = df_titanic_data['Cabin'].map(lambda l: get_cabin_letter(l))
    df_titanic_data['CabinLetter'] = pd.factorize(df_titanic_data['CabinLetter'])[0]

    # binarizing the cabin letters features
    if keep_binary:
        cletters = pd.get_dummies(df_titanic_data['CabinLetter']).rename(columns=lambda x: 'CabinLetter_' + str(x))
        df_titanic_data = pd.concat([df_titanic_data, cletters], axis=1)

    # creating features from the numerical side of the cabin
    df_titanic_data['CabinNumber'] = df_titanic_data['Cabin'].map(lambda x: get_cabin_num(x)).astype(int) + 1
    # scaling the feature
    if keep_scaled:
        scaler_processing = preprocessing.StandardScaler()
# handling the missing values by replacing it with the median feare
    df_titanic_data['Fare'][np.isnan(df_titanic_data['Fare'])] = df_titanic_data['Fare'].median()
        df_titanic_data['CabinNumber_scaled'] =
scaler_processing.fit_transform(df_titanic_data.CabinNumber.reshape(-1, 1))

def get_cabin_letter(cabin_value):
    # searching for the letters in the cabin alphanumerical value
```

```python
    letter_match = re.compile("([a-zA-Z]+)").search(cabin_value)

    if letter_match:
        return letter_match.group()
    else:
        return 'U'

def get_cabin_num(cabin_value):
    # searching for the numbers in the cabin alphanumerical value
    number_match = re.compile("([0-9]+)").search(cabin_value)

    if number_match:
        return number_match.group()
    else:
        return 0

# helper function for constructing features from the ticket fare variable
def process_fare():
    global df_titanic_data

    # handling the missing values by replacing it with the median feare
    df_titanic_data['Fare'][np.isnan(df_titanic_data['Fare'])] = df_titanic_data['Fare'].median()

    # zeros in the fare will cause some division problems so we are going to set them to 1/10th of the lowest fare
    df_titanic_data['Fare'][np.where(df_titanic_data['Fare'] == 0)[0]] = df_titanic_data['Fare'][df_titanic_data['Fare'].nonzero()[0]].min() / 10

    # Binarizing the features by binning them into quantiles
    df_titanic_data['Fare_bin'] = pd.qcut(df_titanic_data['Fare'], 4)
    if keep_binary:
        df_titanic_data = pd.concat([df_titanic_data, pd.get_dummies(df_titanic_data['Fare_bin']).rename(columns=lambda x: 'Fare_' + str(x))], axis=1)

    # binning
    if keep_bins:
        df_titanic_data['Fare_bin_id'] = pd.factorize(df_titanic_data['Fare_bin'])[0] + 1
```

```python
    # scaling the value
    if keep_scaled:
        scaler_processing = preprocessing.StandardScaler()
        df_titanic_data['Fare_scaled'] =
scaler_processing.fit_transform(df_titanic_data.Fare.reshape(-1, 1))

    if keep_bins and keep_scaled:
        scaler_processing = preprocessing.StandardScaler()
        df_titanic_data['Fare_bin_id_scaled'] =
scaler_processing.fit_transform(
            df_titanic_data.Fare_bin_id.reshape(-1, 1))

    if not keep_strings:
        df_titanic_data.drop('Fare_bin', axis=1, inplace=True)

# Helper function for constructing features from the ticket variable
def process_ticket():
    global df_titanic_data

    df_titanic_data['TicketPrefix'] = df_titanic_data['Ticket'].map(lambda
y: get_ticket_prefix(y.upper()))
    df_titanic_data['TicketPrefix'] =
df_titanic_data['TicketPrefix'].map(lambda y: re.sub('[\.?\/?]', '', y))
    df_titanic_data['TicketPrefix'] =
df_titanic_data['TicketPrefix'].map(lambda y: re.sub('STON', 'SOTON', y))

    df_titanic_data['TicketPrefixId'] =
pd.factorize(df_titanic_data['TicketPrefix'])[0]

    # binarzing features for each ticket layer
    if keep_binary:
        prefixes =
pd.get_dummies(df_titanic_data['TicketPrefix']).rename(columns=lambda y:
'TicketPrefix_' + str(y))
        df_titanic_data = pd.concat([df_titanic_data, prefixes], axis=1)
    df_titanic_data.drop(['TicketPrefix'], axis=1, inplace=True)

    df_titanic_data['TicketNumber'] = df_titanic_data['Ticket'].map(lambda
y: get_ticket_num(y))
    df_titanic_data['TicketNumberDigits'] =
df_titanic_data['TicketNumber'].map(lambda y: len(y)).astype(np.int)
    df_titanic_data['TicketNumberStart'] =
df_titanic_data['TicketNumber'].map(lambda y: y[0:1]).astype(np.int)
```

```python
        df_titanic_data['TicketNumber'] =
df_titanic_data.TicketNumber.astype(np.int)

    if keep_scaled:
        scaler_processing = preprocessing.StandardScaler()
        df_titanic_data['TicketNumber_scaled'] =
scaler_processing.fit_transform(
            df_titanic_data.TicketNumber.reshape(-1, 1))

def get_ticket_prefix(ticket_value):
    # searching for the letters in the ticket alphanumerical value
    match_letter = re.compile("([a-zA-Z\.\/]+)").search(ticket_value)
    if match_letter:
        return match_letter.group()
    else:
        return 'U'

def get_ticket_num(ticket_value):
    # searching for the numbers in the ticket alphanumerical value
    match_number = re.compile("([\d]+$)").search(ticket_value)
    if match_number:
        return match_number.group()
    else:
        return '0'

# constructing features from the passenger class variable
def process_PClass():
    global df_titanic_data

    # using the most frequent value(mode) to replace the messing value
    df_titanic_data.Pclass[df_titanic_data.Pclass.isnull()] =
df_titanic_data.Pclass.dropna().mode().values

    # binarizing the features
    if keep_binary:
        df_titanic_data = pd.concat(
            [df_titanic_data,
pd.get_dummies(df_titanic_data['Pclass']).rename(columns=lambda y:
'Pclass_' + str(y))],
            axis=1)

    if keep_scaled:
```

```python
        scaler_preprocessing = preprocessing.StandardScaler()
        df_titanic_data['Pclass_scaled'] =
scaler_preprocessing.fit_transform(df_titanic_data.Pclass.reshape(-1, 1))


# constructing features based on the family variables subh as SibSp and
Parch
def process_family():
    global df_titanic_data

    # ensuring that there's no zeros to use interaction variables
    df_titanic_data['SibSp'] = df_titanic_data['SibSp'] + 1
    df_titanic_data['Parch'] = df_titanic_data['Parch'] + 1

    # scaling
    if keep_scaled:
        scaler_preprocessing = preprocessing.StandardScaler()
        df_titanic_data['SibSp_scaled'] =
scaler_preprocessing.fit_transform(df_titanic_data.SibSp.reshape(-1, 1))
        df_titanic_data['Parch_scaled'] =
scaler_preprocessing.fit_transform(df_titanic_data.Parch.reshape(-1, 1))

    # binarizing all the features
    if keep_binary:
        sibsps_var =
pd.get_dummies(df_titanic_data['SibSp']).rename(columns=lambda y: 'SibSp_'
+ str(y))
        parchs_var =
pd.get_dummies(df_titanic_data['Parch']).rename(columns=lambda y: 'Parch_'
+ str(y))
        df_titanic_data = pd.concat([df_titanic_data, sibsps_var,
parchs_var], axis=1)

# binarzing the sex variable
def process_sex():
    global df_titanic_data
    df_titanic_data['Gender'] = np.where(df_titanic_data['Sex'] == 'male',
1, 0)

# dropping raw original
def process_drops():
    global df_titanic_data
    drops = ['Name', 'Names', 'Title', 'Sex', 'SibSp', 'Parch', 'Pclass',
```

```
'Embarked', \
'Cabin', 'CabinLetter', 'CabinNumber', 'Age', 'Fare',
'Ticket', 'TicketNumber']
    string_drops = ['Title', 'Name', 'Cabin', 'Ticket', 'Sex', 'Ticket',
'TicketNumber']
    if not keep_raw:
        df_titanic_data.drop(drops, axis=1, inplace=True)
    elif not keep_strings:
        df_titanic_data.drop(string_drops, axis=1, inplace=True)

# handling all the feature engineering tasks
def get_titanic_dataset(binary=False, bins=False, scaled=False,
strings=False, raw=True, pca=False, balanced=False):
    global keep_binary, keep_bins, keep_scaled, keep_raw, keep_strings,
df_titanic_data
    keep_binary = binary
    keep_bins = bins
    keep_scaled = scaled
    keep_raw = raw
    keep_strings = strings

    # reading the train and test sets using pandas
    train_data = pd.read_csv('data/train.csv', header=0)
    test_data = pd.read_csv('data/test.csv', header=0)

    # concatenate the train and test set together for doing the overall
feature engineering stuff
    df_titanic_data = pd.concat([train_data, test_data])

    # removing duplicate indices due to coming the train and test set by
re-indexing the data
    df_titanic_data.reset_index(inplace=True)

    # removing the index column the reset_index() function generates
    df_titanic_data.drop('index', axis=1, inplace=True)

    # index the columns to be 1-based index
    df_titanic_data = df_titanic_data.reindex_axis(train_data.columns,
axis=1)

    # processing the titanic raw variables using the helper functions that
```

```
we defined above
    process_cabin()
    process_ticket()
    process_name()
    process_fare()
    process_embarked()
    process_family()
    process_sex()
    process_PClass()
    process_age()
    process_drops()

    # move the survived column to be the first
    columns_list = list(df_titanic_data.columns.values)
    columns_list.remove('Survived')
    new_col_list = list(['Survived'])
    new_col_list.extend(columns_list)
    df_titanic_data = df_titanic_data.reindex(columns=new_col_list)

    print("Starting with", df_titanic_data.columns.size,
          "manually constructing features based on the interaction between
them...\n", df_titanic_data.columns.values)

    # Constructing features manually based on the interaction between the
individual features
    numeric_features = df_titanic_data.loc[:,
                        ['Age_scaled', 'Fare_scaled', 'Pclass_scaled',
'Parch_scaled', 'SibSp_scaled',
                         'Names_scaled', 'CabinNumber_scaled',
'Age_bin_id_scaled', 'Fare_bin_id_scaled']]
    print("\nUsing only numeric features for automated feature
generation:\n", numeric_features.head(10))

    new_fields_count = 0
    for i in range(0, numeric_features.columns.size - 1):
        for j in range(0, numeric_features.columns.size - 1):
            if i <= j:
                name = str(numeric_features.columns.values[i]) + "*" +
str(numeric_features.columns.values[j])
                df_titanic_data = pd.concat(
                    [df_titanic_data, pd.Series(numeric_features.iloc[:, i]
```

```
                * numeric_features.iloc[:, j], name=name)], axis=1)
                new_fields_count += 1
            if i < j:
                name = str(numeric_features.columns.values[i]) + "+" +
str(numeric_features.columns.values[j])
                df_titanic_data = pd.concat(
                    [df_titanic_data, pd.Series(numeric_features.iloc[:, i]
+ numeric_features.iloc[:, j], name=name)], axis=1)
                new_fields_count += 1
            if not i == j:
                name = str(numeric_features.columns.values[i]) + "/" +
str(numeric_features.columns.values[j])
                df_titanic_data = pd.concat(
                    [df_titanic_data, pd.Series(numeric_features.iloc[:, i]
/ numeric_features.iloc[:, j], name=name)], axis=1)
                name = str(numeric_features.columns.values[i]) + "-" +
str(numeric_features.columns.values[j])
                df_titanic_data = pd.concat(
                    [df_titanic_data, pd.Series(numeric_features.iloc[:, i]
- numeric_features.iloc[:, j], name=name)], axis=1)
                new_fields_count += 2

    print("\n", new_fields_count, "new features constructed")

    # using Spearman correlation method to remove the feature that have
high correlation

    # calculating the correlation matrix
    df_titanic_data_cor = df_titanic_data.drop(['Survived', 'PassengerId'],
axis=1).corr(method='spearman')

    # creating a mask that will ignore correlated ones
    mask_ignore = np.ones(df_titanic_data_cor.columns.size) -
np.eye(df_titanic_data_cor.columns.size)
    df_titanic_data_cor = mask_ignore * df_titanic_data_cor

    features_to_drop = []

    # dropping the correclated features
    for column in df_titanic_data_cor.columns.values:
```

```python
        # check if we already decided to drop this variable
        if np.in1d([column], features_to_drop):
            continue

        # finding highly correlacted variables
        corr_vars = df_titanic_data_cor[abs(df_titanic_data_cor[column]) > 0.98].index
        features_to_drop = np.union1d(features_to_drop, corr_vars)

    print("\nWe are going to drop", features_to_drop.shape[0], " which are highly correlated features...\n")
    df_titanic_data.drop(features_to_drop, axis=1, inplace=True)
    # splitting the dataset to train and test and do PCA
    train_data = df_titanic_data[:train_data.shape[0]]
    test_data = df_titanic_data[test_data.shape[0]:]

    if pca:
        print("reducing number of variables...")
        train_data, test_data = reduce(train_data, test_data)
    else:
        # drop the empty 'Survived' column for the test set that was created during set concatenation
        test_data.drop('Survived', axis=1, inplace=True)

    print("\n", train_data.columns.size, "initial features generated...\n")
    # , input_df.columns.values

    return train_data, test_data

# reducing the dimensionality for the training and testing set
def reduce(train_data, test_data):
    # join the full data together
    df_titanic_data = pd.concat([train_data, test_data])
    df_titanic_data.reset_index(inplace=True)
    df_titanic_data.drop('index', axis=1, inplace=True)
    df_titanic_data = df_titanic_data.reindex_axis(train_data.columns, axis=1)

    # converting the survived column to series
    survived_series = pd.Series(df_titanic_data['Survived'], name='Survived')
```

```python
    print(df_titanic_data.head())

    # getting the input and target values
    input_values = df_titanic_data.values[:, 1::]
    target_values = df_titanic_data.values[:, 0]

    print(input_values[0:10])

    # minimum variance percentage that should be covered by the reduced
number of variables
    variance_percentage = .99

    # creating PCA object
    pca_object = PCA(n_components=variance_percentage)

    # trasforming the features
    input_values_transformed = pca_object.fit_transform(input_values,
target_values)

    # creating a datafram for the transformed variables from PCA
    pca_df = pd.DataFrame(input_values_transformed)

    print(pca_df.shape[1], " reduced components which describe ",
str(variance_percentage)[1:], "% of the variance")

    # constructing a new dataframe that contains the newly reduced vars of
PCA
    df_titanic_data = pd.concat([survived_series, pca_df], axis=1)

    # split into separate input and test sets again
    train_data = df_titanic_data[:train_data.shape[0]]
    test_data = df_titanic_data[test_data.shape[0]:]
    test_data.reset_index(inplace=True)
    test_data.drop('index', axis=1, inplace=True)
    test_data.drop('Survived', axis=1, inplace=True)

    return train_data, test_data

# Calling the helper functions
if __name__ == '__main__':
```

```
train, test = get_titanic_dataset(bins=True, scaled=True, binary=True)
initial_drops = ['PassengerId']
train.drop(initial_drops, axis=1, inplace=True)
test.drop(initial_drops, axis=1, inplace=True)

train, test = reduce(train, test)

print(train.columns.values)
```

3.4 偏差-方差分解

在前文中，我们已经知道了如何选择模型的最优超参数。这组最优超参数是基于最小化交叉验证误差选出来的。现在，我们需要了解模型在未知的数据或者所谓的样本外数据上表现如何，样本外数据是指在模型训练阶段完全没有看到过的新数据样本。

考虑下面这个例子：假如有一个大小为 10000 的数据样本，这里使用不同大小的训练集来训练同样的模型，并绘制每一步的测试误差。例如，我们取出其中 1000 个样本作为测试集，使用剩余的 9000 个样本来训练。所以，对于第一轮训练，将从 9000 个样本中随机抽取一个大小为 100 的训练集。然后基于之前选出的最优超参数组训练模型，并使用测试集来测试模型，接着画出训练（样本内）误差和测试（样本外）误差。最后使用不同的训练集大小（例如，重复从 9000 个样本中选 500 个样本，然后从 9000 个样本中选 1000 个样本等）重复上述训练、测试、画图操作。

在完成所有这些训练、测试和画图后，我们将得到一个包含两条曲线的图，代表使用相同模型但训练集大小不同的训练误差和测试误差。从这幅图中，我们可以了解到模型到底表现怎样。

这个包含分别代表训练误差和测试误差的两条曲线的输出图可能会呈现出图 3.2 所示的 4 种形状之一。这 4 种不同形状的曲线来自 Andrew Ng 在 Coursera 网站上的机器学习课程。这是一门优秀课程，对于机器学习新手来说，其中有很多很好的见解和实践。

所以，什么时候才可以接受模型并投入使用？何时能够知道模型在测试集上表现不佳，并且不会出现错误的泛化误差？问题的答案取决于在不同大小训练集上所绘制的训练误差和测试误差关系曲线的形状。

3.4 偏差-方差分解

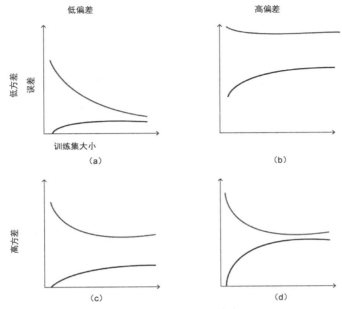

图 3.2 在不同大小训练集上绘制训练误差和测试误差可能出现的曲线形状

- 如果曲线形状看起来像图 3.2（a）中的图形，那么它代表了低的训练误差并且在测试集上泛化效果良好。这种模型是好的，当遇到这种模型的时候应该继续往前，并可以投入实际使用中。

- 如果曲线形状看起来像图 3.2（b）中的图形，那么它代表了高的训练误差（模型没有从训练样本中学习到什么），并且在测试集上泛化效果更差。这种曲线完全就是失败的，如果遇到这种情况需要读者回过头来看看数据、选择的学习算法、选择的超参数有什么问题。

- 如果曲线形状看起来像图 3.2（c）中的图形，它代表了不好的训练误差，因为模型没有能够捕捉到数据的基本结构，这同样适用于新的测试数据。

- 如果曲线形状看起来像图 3.2（d）中的图形，它代表了很高的偏差和方差。这意味着模型没有很好地理解训练数据，因此在测试集上泛化效果也不佳。

偏差和方差是可以用来确定模型表现如何的指标。在监督学习中，存在两种相反的误差来源。使用图 3.2 中的学习曲线，可以弄清楚模型中究竟哪里出了问题。具有高方差和低偏差的问题通常称为**过拟合**，这意味着模型在训练样本上表现很好，但是在测试集上泛化效果不佳。具有高偏差和低方差的问题称为**欠拟合**，这意味着模型没有充分利用数据，并且没有设法从输入特征中估计输出（目标）。有很多不同的方法可以用来避免出现二者中的任何一种问题。然而，通常，优化其中一种指标将以牺牲另一种指标为代价。

对于高方差的情况，一般通过增加模型可以从中学习的更多特征来解决。这些解决方法很有可能增加偏差，所以需要考虑在两者之间做一些权衡。

3.5 学习可见性[①]

有很多优秀的数据科学算法能够用来解决不同领域的问题，但是使学习过程可见的关键问题在于拥有足够的数据。读者可能会问，学习过程到底需要多少数据才能有效并且值得做。根据经验，研究人员和机器学习从业者认为数据样本的大小至少需要是模型**自由度**的 10 倍。

例如，对于线性模型，自由度表示数据集中特征的数目。如果数据集中有 50 个解释性特征，那么至少需要 500 个数据样本（观测值）。

打破经验法则

在实践中，我们可以打破经验法则规则，在数据样本小于数据特征数的 10 倍时仍然进行学习。这通常发生在模型比较简单或者使用了类似于**正则化**（在下一章会讨论）的方法的情况下。

Jake Vanderplas 写了一篇文章来证明即使数据拥有比样本更多的参数仍然可以学习。为了证明这一点，他使用了正则化。

3.6 总结

本章介绍了机器学习从业者使用的最重要工具，目的是能够理解数据，并且使得学习算法能够最大限度地利用好这些数据。

特征工程是数据科学中第一个很常用的工具，是任何数据科学流程中必须要有的组件。这个工具的目的就是能够更好地表征数据，然后增强模型的预测能力。

[①] 此处原文为"Learning visibility"，译者认为意思是"学习的过程有明显的学习效果"，根据后文来理解为"需要多少数据才能明显地看到正在学习并且学习到了东西"。——译者注

从前面的介绍可以看出，大量的特征是有问题的，并且可能会导致更差的分类器性能。同时，存在一个最优的特征数目，它能够使得模型有最好的性能，并且这个最优特征数目是获得的数据样本（观测数量）的函数。

随后，本章介绍了一种非常强大的工具，那就是偏差-方差分解。这个工具普遍用于测试模型在测试集上表现的好坏。

最后，我们梳理了学习可见性，它回答了究竟需要多少数据来开展业务和使用机器学习的问题。经验表明，通常至少需要 10 倍于数据特征数的数据样本/观测。然而，如果使用了另一种名为正则化的工具，这种规则可能被打破，下一章会更详细地讨论正则化工具。

接下来，我们会继续增加能够从数据中得到有意义结论的数据科学工具，然后会讨论一些应用机器学习的日常问题。

第 4 章
TensorFlow 入门实战

本章介绍一个现阶段使用最广泛的深层学习框架——TensorFlow。因为 TensorFlow 所拥有的社区支持度日益增长,所以使用它来构建复杂的深度学习应用程序是我们的不二选择。从 TensorFlow 网站我们可以看到:

> TensorFlow 是一个使用数据流图进行数值计算的开源软件库。图中的节点代表数学运算,而图中的边则代表在这些节点之间传递的多维数组(张量)。借助这种灵活的架构,你可以通过一个 API 将计算工作部署到台式机、服务器或移动设备中的一个或多个 CPU 或 GPU 上。TensorFlow 最初是由 Google Brain 团队(隶属于 Google 机器智能研究部门)的研究人员和工程师开发的,旨在用于进行机器学习和深度神经网络研究。该系统具有很好的通用性,还可以应用于众多其他领域。

本章将介绍以下内容。

- 安装 TensorFlow。
- TensorFlow 运行环境。
- 计算图。
- TensorFlow 中的数据类型、变量和占位符。
- 获取 TensorFlow 的输出。
- TensorBoard——可视化学习过程。

4.1 安装 TensorFlow

TensorFlow 有两种版本可以安装,分别是 CPU 版和 GPU 版。本书将使用 GPU 版。

4.1.1 在 Ubuntu 16.04 系统上安装 GPU 版的 TensorFlow

在安装 GPU 版的 TensorFlow 之前，首先需要安装最新版的 NVIDIA 驱动程序，因为现阶段 GPU 版的 TensorFlow 只支持 CUDA。接下来将会带领读者一步步地安装 NVIDIA 驱动程序和 CUDA8 驱动程序。

1. 安装 NVIDIA 驱动程序和 CUDA 8

首先需要安装正确的 NVIDIA 驱动程序。这里以 GeForce GTX 960M GPU 为例，因此会安装 nvidia-375（如果读者使用的是其他版本，请通过 NVIDIA 官网来找到适合你的驱动程序）。如果你想确认你的 GPU 型号，可以在终端中使用如下命令。

```
lspci | grep -i nvidia
```

你将会看到以下输出。

接下来，需要添加一个专有的 NVIDIA 驱动程序库，以便能够使用 `apt-get` 安装驱动程序。

```
sudo add-apt-repository ppa:graphics-drivers/ppa
sudo apt-get update
sudo apt-get install nvidia-375
```

在成功安装 NVIDIA 驱动程序之后，重启计算机，在终端输入以下命令来确认是否成功安装驱动程序。

```
cat /proc/driver/nvidia/version
```

终端的输出结果如下。

接下来，安装 CUDA 8。打开 NVIDIA 官网的 CUDA 下载链接，根据图 4.1 所示的屏幕截图来选择你的操作系统、体系结构、发行版、版本号以及安装程序类型。

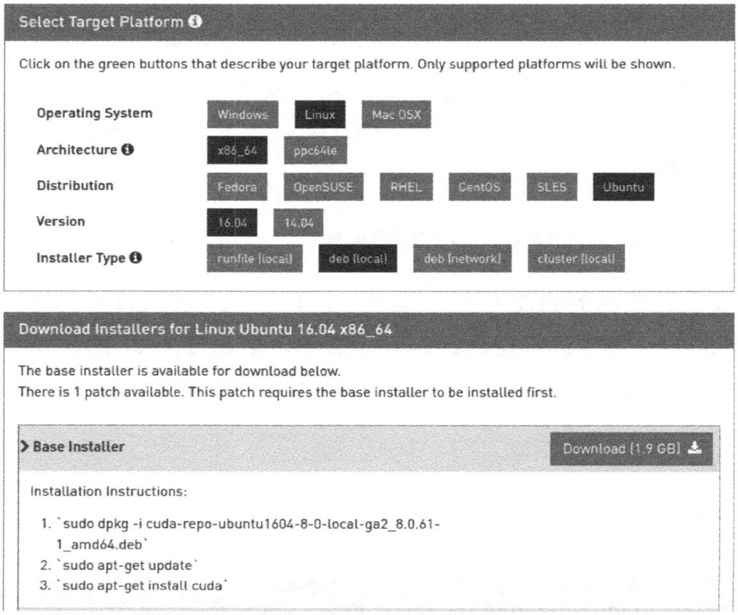

图 4.1 CUDA 8 安装过程屏幕截图

安装程序大约为 2GB。需要使用如下安装命令。

```
sudo dpkg -i cuda-repo-ubuntu1604-8-0-local-ga2_8.0.61-1_amd64.deb
sudo apt-get update
sudo apt-get install cuda
```

接下来，需要通过以下命令将这些库添加到 .bashrc 文件中。

```
echo 'export PATH=/usr/local/cuda/bin:$PATH' >> ~/.bashrc
echo 'export LD_LIBRARY_PATH=/usr/local/cuda/lib64:$LD_LIBRARY_PATH' >> ~/.bashrc
source ~/.bashrc
```

接下来，需要通过以下命令来验证 CUDA 8 是否成功安装。

```
nvcc -V
```

如果安装成功，将在终端中看到以下结果。

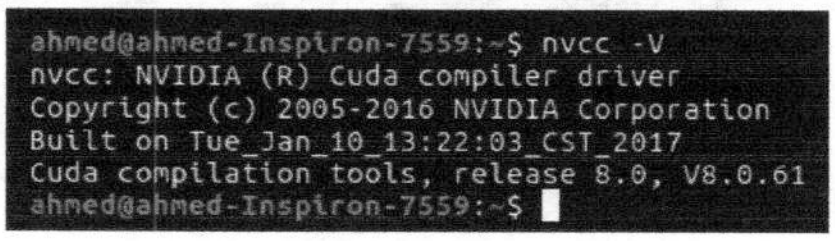

最后，安装 cuDNN 6.0。**NVIDIA CUDA 深度神经网络库**（NVIDIA CUDA Deep Neural Network library，cuDNN）是一个支持 GPU 加速的深层神经网络库。可以从 NVIDIA 官网下载它。使用以下命令来提取和安装 cuDNN。

```
cd ~/Downloads/
tar xvfcudnn*.tgz
cd cuda
sudo cp */*.h /usr/local/cuda/include/
sudo cp */libcudnn* /usr/local/cuda/lib64/
sudo chmod a+r /usr/local/cuda/lib64/libcudnn*
```

为确保安装成功，可以在终端中使用 nvidia-smi 工具来进行验证。如果安装成功，该工具将提供 GPU 的监控信息，例如 RAM 和 GPU 的运行状态。

2. 安装 TensorFlow

在为 TensorFlow 准备好 GPU 环境之后，现在可以在 GPU 模式下安装 TensorFlow 了。在安装 TensorFlow 之前，可以先安装一些有用的 Python 软件包。这些包将在下一章中讲解，它们可以使开发环境使用起来更加简单。

通过以下命令开始安装一些有关数据操作、分析和可视化的库。

```
sudo apt-get update && apt-get install -y python-numpy python-scipy python-nose python-h5py python-skimage python-matplotlib python-pandas python-sklearn python-sympy
sudo apt-get clean && sudo apt-get autoremove
sudo rm -rf /var/lib/apt/lists/*
```

也可以安装更多有用的工具库，比如虚拟环境（virtual environment）、Jupyter、Notebook 等。

```
sudo apt-get update
sudo apt-get install git python-dev python3-dev python-numpy python3-numpy build-essential python-pip python3-pip python-virtualenv swig python-wheel libcurl3-dev
sudo apt-get install -y libfreetype6-dev libpng12-dev
pip3 install -U matplotlibipython[all] jupyter pandas scikit-image
```

最后，通过使用以下命令安装 TensorFlow 的 GPU 版。

```
pip3 install --upgrade tensorflow-gpu
```

接下来，使用 Python 来验证 TensorFlow 是否安装成功。

```
python3
>>> import tensorflow as tf
>>> a = tf.constant(5)
>>> b = tf.constant(6)
>>> sess = tf.Session()
>>> sess.run(a+b)
// this should print bunch of messages showing device status etc. // If everything
goes well, you should see gpu listed in device
>>> sess.close()
```

如果安装成功将会在终端中看到以下内容。

4.1.2　在 Ubuntu 16.04 系统上安装 CPU 版的 TensorFlow

在本节中，你将会安装 CPU 版的 TensorFlow，它的好处是在安装之前不需要安装额外的驱动程序。在安装开始之前建议你先安装一些有关数据操作和可视化的软件包。

```
sudo apt-get update && apt-get install -y python-numpy python-scipy python-nose
python-h5py python-skimage python-matplotlib python-pandas python-sklearn python-sympy
sudo apt-get clean && sudo apt-get autoremove
sudo rm -rf /var/lib/apt/lists/*
```

4.1 安装 TensorFlow

也可以安装更多有用的工具库，比如虚拟环境（virtual environment）、Jupyter、Notebook 等。

```
sudo apt-get update
sudo apt-get install git python-dev python3-dev python-numpy python3-numpy
build-essential  python-pip python3-pip python-virtualenv swig python-wheel
libcurl3-dev
sudo apt-get install -y libfreetype6-dev libpng12-dev
pip3 install -U matplotlibipython[all] jupyter pandas scikit-image
```

最后，安装最新的 CPU 版的 TensorFlow。

```
pip3 install --upgrade tensorflow
```

接下来，为了验证是否安装成功，可以尝试以下命令。

```
python3
>>> import tensorflow as tf
>>> a = tf.constant(5)
>>> b = tf.constant(6)
>>> sess = tf.Session()
>>> sess.run(a+b)
>> sess.close()
```

如果安装成功，将会在终端中看到以下结果。

4.1.3　在 Mac OS X 上安装 CPU 版的 TensorFlow

本节将介绍使用 virtualenv 为 Mac OS X 安装 TensorFlow。首先，通过以下命令来安装 pip 工具。

```
sudo easy_install pip
```

接下来，安装虚拟环境库。

```
sudo pip install --upgrade virtualenv
```

安装完虚拟环境库之后，我们需要创建一个容器或虚拟环境来承载安装的 TensorFlow，以及可能要安装的任何软件包，但不会影响底层主机系统。

```
virtualenv --system-site-packages targetDirectory # for Python 2.7
virtualenv --system-site-packages -p python3 targetDirectory # for Python 3.n
```

这里所指的 targetDirectory 是 ~/tensorflow。

现在虚拟环境已经创建好了，使用以下命令访问这个环境。

```
source ~/tensorflow/bin/activate
```

一旦你使用这个命令，就可以访问刚刚创建的虚拟环境了，并且可以安装任何只使用在这个环境中的软件包，而不会对底层或主机系统产生影响。

要退出虚拟环境，可以使用以下命令。

```
Deactivate
```

要进入虚拟环境，可以使用以下命令。一旦使用完 TensorFlow，就应该关闭它。

```
source bin/activate
```

为了安装 CPU 版的 TensorFlow，可以使用以下命令，该命令还将安装 TensorFlow 所需的任何相关库。

```
(tensorflow)$ pip install --upgrade tensorflow      # for Python 2.7
(tensorflow)$ pip3 install --upgrade tensorflow     # for Python 3.n
```

4.1.4　在 Windows 系统上安装 CPU/GPU 版的 TensorFlow

假定你的系统上已经安装了 Python 3。要安装 TensorFlow，请按如下操作步骤以管理员身份启动命令提示符。具体做法为：打开"开始"菜单，搜索"cmd"，然后右击它，并选择"以管理员身份运行"，如图 4.2 所示。

图 4.2 开始安装 TensorFlow

一旦打开了一个命令窗口，即可以使用以下命令在 GPU 模式下安装 TensorFlow。

```
C:\> pip3 install --upgrade tensorflow-gpu
```

 在使用下一个命令之前，读者需要安装 pip 或 pip3（根据 Python 版本）。

或者使用以下命令安装 CPU 版的 TensorFlow。

```
C:\> pip3 install --upgrade tensorflow
```

4.2 TensorFlow 运行环境

TensorFlow 是谷歌的一个深度学习框架，顾名思义，它来源于神经网络在多维数据阵列或张量（Tensor）上执行的操作。TensorFlow 实际上是张量流。

但首先我们要明确为什么要在本书中使用深度学习框架。

- **它简化了机器学习代码**。由于这些深度学习框架的存在，大部分深度学习和机器学习的研究都可以实现。它们允许数据科学家极其快速地迭代自己的深度学习项目，并为从业者提供了更多使用深度学习和其他机器学习算法的机会。谷歌、Facebook 等大公司正在使用这种深度学习框架来扩展更多的用户。
- **它可以计算梯度**。深度学习框架也可以自动计算梯度。如果尝试过逐步进行梯度计算，我们会发现梯度计算并不简单，而且自己编写出无 Bug 的程序可能会非常困难。
- **它将机器学习应用程序标准化以进行共享**。另外，可以在不同的深度学习框架中使用的预训练模型可以在线获得。这些预训练模型可以帮助那些 GPU 资源有限的人，这样他们就不必每次都从头开始。我们可以站在巨人的肩膀上了解这些。
- **不同的深度学习框架具有不同的优点**。例如，范例、抽象层次、编程语言等。
- **它会提供 GPU 并行处理的接口**。使用 GPU 进行计算是一项非常吸引人的功能，因为由于庞大的内核数量和并行化，GPU 加速代码的速度比 CPU 快得多。

这就是为什么想在深度学习方面取得突破，TensorFlow 是读者的必然选择，因为它可以为读者的项目提供便利。

那么，简单地说，什么是 TensorFlow？

- TensorFlow 是来自谷歌的深度学习框架，它是使用数据流图进行数值计算的开源软件。
- TensorFlow 最初是由谷歌 Brain Team 开发的，目的是促进他们的机器学习研究。
- TensorFlow 可以理解为机器学习算法的编程接口和执行这些算法的实现工具。

那么 TensorFlow 是怎么工作的？它的规则是什么？

4.3 计算图

关于 TensorFlow 的所有重要创新中最著名的是：把所有数字计算都表示为计算图，如图 4.3 所示。任何 TensorFlow 程序都将变成一个计算图，具体细节如下。

- 图中的节点是有关任意输入和输出的运算节点。
- 图中的节点之间的边可以理解为在运算之间流动的张

图 4.3 TensorFlow 计算图

量。简言之，在实践中张量就是 n 维数组。

使用图 4.3 所示的计算图作为深度学习框架的核心，其优势在于它允许我们根据小而简单的操作构建复杂的模型。另外，在后面的章节中会发现，这将使梯度计算变得非常简单。

理解 TensorFlow 计算图概念的思路是，每个运算都是一个可以在图节点上进行计算的函数。

4.4 TensorFlow 中的数据类型、变量、占位符

对计算图的理解将帮助我们根据小的子图和运算来思考复杂的模型。

看一个只有一个隐藏层的神经网络的例子，以及在 TensorFlow 中它的计算图的形状。

$$h = \text{ReLU}(Wx + b)$$

在这里求解经过 ReLU 激活后的结果，首先某个参数矩阵 W 乘以某个输入 x 再加上一个偏差项 b，然后使用 ReLU 函数，取输出和零中的最大值。

在图 4.3 所示计算图中，有变量 b、矩阵 W 和一个叫作 x 的占位符，图中同时包含每个运算的节点。下面进一步讲述这些节点类型。

4.4.1 变量

变量可以作为有状态的节点从而输出其当前值。在这个例子中，变量是 b 和 W。变量是有状态的含义是在多次执行之后，程序依然保留了它们的当前值，并且很容易将已保存的值还原为变量，如图 4.4 所示。

另外，变量还有其他有用的功能。例如，变量可以在训练期间或者训练后保存到硬盘中，这有助于实现前面提到的功能：可以让不同公司与团队的人员保存、存储模型参数并将它们发送给其他人。此外，变量可以调整变量以减少损失（loss），接下来我们将看到如何做到这一点。

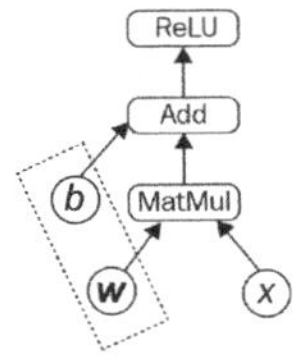

图 4.4 TensorFlow 计算图

注意，计算图中的变量（如 b 和 W）仍然表示运算，因为根据定义，图中的所有节点都表示操作。因此，在运行过程中当求 b 和 W 的值时，我们将获得这些变量的值。

可以使用 TensorFlow 的 Variable()函数来定义一个变量并给它赋初始值。

```
var = tf.Variable(tf.random_normal((0,1)),name='random_values')
```

这行代码将定义一个 1×1 的变量并使用标准正态分布初始化它[①]，同时也可以给这个变量命名。

4.4.2 占位符

下一个类型的节点是占位符。占位符是指其值在执行时传入的节点，如图 4.5 所示。

如果计算图中有依赖于某些外部数据的输入，那么可以使用占位符在训练期间添加值到计算图中。所以，对于占位符，我们不需要提供任何初始值，只需要给张量分配数据类型和形状。这样计算图就知道要计算什么，即使它尚未存储任何值。

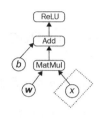

图 4.5 TensorFlow 计算图

可以使用 TensorFlow 的 placeholder()函数来创建占位符。

```
ph_var1 = tf.placeholder(tf.float32,shape=(2,3))
ph_var2 = tf.placeholder(tf.float32,shape=(3,2))
result = tf.matmul(ph_var1,ph_var2)
```

这些代码行定义了两个特定形状的占位符变量，然后定义了将这两个值相乘的运算（具体介绍请参考 4.4.3 节）。

4.4.3 数学运算

第三类节点是数学运算，包括矩阵乘法（MatMul）、加法（Add）和 ReLU，如图 4.6 所示。所有这些运算都是 TensorFlow 图中的节点，Tensorflow 中的这些操作与 NumPy 操作非常相似。

下面讨论计算图在代码中的形状。

执行以下步骤来生成图 4.6。

图 4.6 TensorFlow 计算图

1）创建并初始化权重 W 和 b。初始化权重。矩阵 W 的方法是从均匀分布 W～Uniform $(-1,1)$ 采样并且初始化 b 为 0。

[①] 原文为"定义一个 2×2 的变量"，应为笔误。——译者注

2）创建输入占位符 x，输入具有 $m*784$ 形状的矩阵。

3）建立一个流程图。

继续并按照以下步骤来构建流程图。

```
# import TensorFlow package
import tensorflow as tf
# build a TensorFlow variable b taking in initial zeros of size 100
# ( a vector of 100 values)
b = tf.Variable(tf.zeros((100,)))
# TensorFlow variable uniformly distributed values between -1 and 1
# of shape 784 by 100
W = tf.Variable(tf.random_uniform((784, 100),-1,1))
# TensorFlow placeholder for our input data that doesn't take in
# any initial values, it just takes a data type 32 bit floats as
# well as its shape
x = tf.placeholder(tf.float32, (100, 784))
# express h as TensorflowReLU of the TensorFlow matrix
#Multiplication of x and W and we add b
h = tf.nn.relu(tf.matmul(x,W) + b )
```

从前面的代码可以看出，我们实际上并没有用这段代码来操纵任何数据。我们只是在图形内部构建符号，并且在运行此图表之前，无法输出 h 并查看其值。所以，这段代码只用于构建模型的核心。如果你尝试在前面的代码中输出 W 或 b 的值，则应该在 Python 中执行以下操作。

```
ahmed@ahmed-Inspiron-7559:~$ python3
Python 3.6.0 |Anaconda 4.3.0 (64-bit)| (default, Dec 23 2016, 12:22:00)
[GCC 4.4.7 20120313 (Red Hat 4.4.7-1)] on linux
Type "help", "copyright", "credits" or "license" for more information.
>>> # import TensorFlow package
... import tensorflow as tf
>>> # build a TensorFlow variable b taking in initial zeros of size 100
... # ( a vector of 100 values)
... b = tf.Variable(tf.zeros((100,)))
>>> # TensorFlow variable uniformly distributed values between -1 and 1
... # of shape 784 by 100
... W = tf.Variable(tf.random_uniform((784, 100),-1,1))
>>> # TensorFlow placeholder for our input data that doesn't take in
... # any initial values, it just takes a data type 32 bit floats as
... # well as its shape
... x = tf.placeholder(tf.float32, (100, 784))
>>> # express h as Tensorflow ReLU of the TensorFlow matrix
... #Multiplication of x and W and we add b
... h = tf.nn.relu(tf.matmul(x,W) + b )
>>> print(W)
<tf.Variable 'Variable_1:0' shape=(784, 100) dtype=float32_ref>
>>> print(b)
<tf.Variable 'Variable:0' shape=(100,) dtype=float32_ref>
>>>
```

第 4 章　TensorFlow 入门实战

至此，我们已经定义了图形，现在需要实际运行它。

4.5　获取 TensorFlow 的输出

在 4.4 节中，读者知道了如何构建计算图，现在介绍如何实际运行它并获得它的输出。

我们可以用会话（session）来部署/运行计算图，会话可以将计算任务部署到一个特定的运算环境（如 CPU 或 GPU）中。因此，可以使用会话将构建的计算图部署到 CPU 或 GPU 中。

为了运行计算图，需要定义一个名为 sess 的会话对象，并且调用带有两个参数的函数 run。

```
sess.run(fetches, feeds)
```

其中，fetches 指的是需要输出的所有计算图节点构成的列表，可以理解为我们希望计算的关键节点；feeds 指的是从图中的节点到读者想要在模型中运行的实际值的字典映射，因此，feeds 就是之前提到的要传入数值到占位符的地方。

详细代码如下所示。

```
# importing the numpy package for generating random variables for
# our placeholder x
import numpy as np
# build a TensorFlow session object which takes a default execution
# environment which will be most likely a CPU
sess = tf.Session()
# calling the run function of the sess object to initialize all the
# variables.
sess.run(tf.global_variables_initializer())
# calling the run function on the node that we are interested in,
# the h, and we feed in our second argument which is a dictionary
# for our placeholder x with the values that we are interested in.
sess.run(h, {x: np.random.random((100,784))})
```

在通过 sess 对象运行计算图后，应该得到类似于以下的输出。

```
>>> # importing the numpy package for generating random variables for
... # our placeholder x
... import numpy as np
>>> # build a TensorFlow session object which takes a default execution
... # environment which will be most likely a CPU
... sess = tf.Session()
>>> # calling the run function of the sess object to initialize all the
... # variables.
... sess.run(tf.global_variables_initializer())
>>> # calling the run function on the node that we are interested in,
... # the h, and we feed in our second argument which is a dictionary
... # for our placeholder x with the values that we are interested in.
... sess.run(h, {x: np.random.random((100,784))})
array([[ 4.95583916,  0.13156724,  0.        , ...,  0.        ,
         0.        ,  0.        ],
       [ 0.        ,  0.        ,  0.        , ...,  0.        ,
         0.        ,  0.        ],
       [ 0.        ,  0.        ,  0.        , ...,  0.        ,
         0.        ,  0.        ],
       ...,
       [ 2.81067681,  8.28696823,  0.        , ...,  0.        ,
         0.55022001,  0.        ],
       [ 0.        ,  6.67730427,  0.        , ...,  0.        ,
         4.28411198,  3.10559845],
       [ 1.92718267,  0.        ,  0.        , ...,  0.        ,
         0.        ,  0.        ]], dtype=float32)
>>>
```

可以看到，在上面的代码片段的第 9 行[1]中，初始化变量，这是 TensorFlow 中的一个概念，它称为延后计算（lazy evaluation）。这意味着计算图只会在会话运行时计算。所以，调用函数 global_variables_initializer()实际上会初始化计算图中所有的变量，比如之前例子中的 W 和 b。

我们还可以在 with 代码块中使用会话变量，以确保在执行完该计算图后会话将关闭。

```
ph_var1 = tf.placeholder(tf.float32,shape=(2,3))
ph_var2 = tf.placeholder(tf.float32,shape=(3,2))
result = tf.matmul(ph_var1,ph_var2)
with tf.Session() as sess:
print(sess.run([result],feed_dict={ph_var1:[[1.,3.,4.],[1.,3.,4.]],ph_var2:
[[1., 3.],[3.,1.],[.1,4.]]}))


Output:
[array([[10.4, 22. ],
       [10.4, 22. ]], dtype=float32)]
```

4.6　TensorBoard——可视化学习过程

当我们使用 TensorFlow 训练一个大规模的深度神经网络时，过程可能会很复杂且令人困惑，同时其相应的计算图也会很复杂。为了更容易地理解训练过程、调试和优化

---

[1] 原文此处为"第 2 行"，应为笔误。——译者注

TensorFlow 程序，TensorFlow 团队开发了一套名为 TensorBoard 的可视化工具，它是一套可以通过浏览器运行的 Web 应用程序。TensorBoard 可用于可视化 TensorFlow 计算图，绘制有关计算图运行结果的量化指标，并显示其他数据（如通过它的图像）。配置好的 TensorBoard 界面如图 4.7 所示。

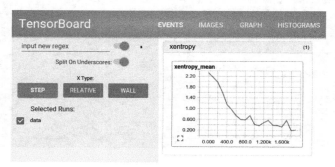

图 4.7 配置好的 TensorBoard 界面

为了理解 TensorBoard 是如何工作的，我们使用由 MNIST 数据集训练好的分类器构成的计算图作为例子，MNIST 是手写图像的数据集。

读者不需要了解这个模型的所有部分，TensorBoard 会展示在 TensorFlow 中实现的机器学习模型的流程图。

因此，我们首先导入 TensorFlow，并使用 TensorFlow 帮助函数加载所需的数据集。这些帮助函数将检查你是否已经下载了数据集，如果未下载，那么它会自动下载。

```
import tensorflow as tf

# Using TensorFlow helper function to get the MNIST dataset from tensorflow.examples.
tutorials.mnist import input_data
mnist_dataset = input_data.read_data_sets("/tmp/data/", one_hot=True)

Output:
Extracting /tmp/data/train-images-idx3-ubyte.gz
Extracting /tmp/data/train-labels-idx1-ubyte.gz
Extracting /tmp/data/t10k-images-idx3-ubyte.gz
Extracting /tmp/data/t10k-labels-idx1-ubyte.gz
```

接下来，定义超参数（可用于微调模型性能的参数）和模型的输入。

```
# hyperparameters of the the model (you don't have to understand the functionality
of each parameter)
```

```python
learning_rate = 0.01
num_training_epochs = 25
train_batch_size = 100
display_epoch = 1
logs_path = '/tmp/tensorflow_tensorboard/'

# Define the computational graph input which will be a vector of the image pixels
# Images of MNIST has dimensions of 28 by 28 which will multiply to 784
input_values = tf.placeholder(tf.float32, [None, 784], name='input_values')

# Define the target of the model which will be a classification problem of 10 classes
from 0 to 9
target_values = tf.placeholder(tf.float32, [None, 10], name='target_values')

# Define some variables for the weights and biases of the model
weights = tf.Variable(tf.zeros([784, 10]), name='weights')
biases = tf.Variable(tf.zeros([10]), name='biases')
```

现在需要建立模型并定义一个将要优化的损失函数。

```python
# Create the computational graph and encapsulating different operations to different scopes
# which will make it easier for us to understand the visualizations of TensorBoard
with tf.name_scope('Model'):
    # Defining the model
    predicted_values = tf.nn.softmax(tf.matmul(input_values, weights) + biases)

with tf.name_scope('Loss'):
# Minimizing the model error using cross entropy criteria
model_cost = tf.reduce_mean(-
tf.reduce_sum(target_values*tf.log(predicted_values), reduction_indices=1))

with tf.name_scope('SGD'):
# using Gradient Descent as an optimization method for the model cost above
model_optimizer =
tf.train.GradientDescentOptimizer(learning_rate).minimize(model_cost)

with tf.name_scope('Accuracy'):
#Calculating the accuracy
model_accuracy = tf.equal(tf.argmax(predicted_values, 1),
tf.argmax(target_values, 1))
model_accuracy = tf.reduce_mean(tf.cast(model_accuracy, tf.float32))
```

```
# TensorFlow use the lazy evaluation strategy while defining the variables
# So actually till now none of the above variable got created or
initialized
init = tf.global_variables_initializer()
```

定义 summary 变量,用于监视特定变量发生的变化,如损失值以及整个训练过程中的拟合情况。

```
# Create a summary to monitor the model cost tensor
tf.summary.scalar("model loss", model_cost)

# Create another summary to monitor the model accuracy tensor
tf.summary.scalar("model accuracy", model_accuracy)

# Merging the summaries to single operation
merged_summary_operation = tf.summary.merge_all()
```

最后,我们将通过定义一个会话变量来运行模型,该变量将用于执行刚刚构建的计算图。

```
# kick off the training process
with tf.Session() as sess:

 # Intialize the variables
 sess.run(init)

 # operation to feed logs to TensorBoard
 summary_writer = tf.summary.FileWriter(logs_path,
graph=tf.get_default_graph())

 # Starting the training cycle by feeding the model by batch at a time
 for train_epoch in range(num_training_epochs):

 average_cost = 0.
 total_num_batch = int(mnist_dataset.train.num_examples/train_batch_size)
 # iterate through all training batches
 for i in range(total_num_batch):
 batch_xs, batch_ys = mnist_dataset.train.next_batch(train_batch_size)

 # Run the optimizer with gradient descent and cost to get the loss
 # and the merged summary operations for the TensorBoard
 _, c, summary = sess.run([model_optimizer, model_cost,
merged_summary_operation],
 feed_dict={input_values: batch_xs, target_values: batch_ys})
```

4.6 TensorBoard——可视化学习过程

```
            # write statistics to the log et every iteration
            summary_writer.add_summary(summary, train_epoch * total_num_batch + i)

            # computing average loss
            average_cost += c / total_num_batch

        # Display logs per epoch step
        if (train_epoch+1) % display_epoch == 0:
            print("Epoch:", '%03d' % (train_epoch+1), "cost=", "{:.9f}".format(average_cost))

    print("Optimization Finished!")

    # Testing the trained model on the test set and getting the accuracy compared to the actual labels of the test set
    print("Accuracy:", model_accuracy.eval({input_values:
mnist_dataset.test.images, target_values: mnist_dataset.test.labels}))

    print("To view summaries in the Tensorboard, run the command line:\n" \
    "-->tensorboard --logdir=/tmp/tensorflow_tensorboard " \
    "\nThen open http://0.0.0.0:6006/ into your web browser")
```

训练过程的输出应该与以下输出类似。

```
Epoch: 001 cost= 1.183109128
Epoch: 002 cost= 0.665210275
Epoch: 003 cost= 0.552693334
Epoch: 004 cost= 0.498636444
Epoch: 005 cost= 0.465516675
Epoch: 006 cost= 0.442618381
Epoch: 007 cost= 0.425522513
Epoch: 008 cost= 0.412194222
Epoch: 009 cost= 0.401408134
Epoch: 010 cost= 0.392437336
Epoch: 011 cost= 0.384816745
Epoch: 012 cost= 0.378183398
Epoch: 013 cost= 0.372455584
Epoch: 014 cost= 0.367275238
Epoch: 015 cost= 0.362772711
Epoch: 016 cost= 0.358591895
Epoch: 017 cost= 0.354892231
Epoch: 018 cost= 0.351451424
Epoch: 019 cost= 0.348337946
Epoch: 020 cost= 0.345453095
Epoch: 021 cost= 0.342769080
```

```
Epoch: 022 cost= 0.340236065
Epoch: 023 cost= 0.337953151
Epoch: 024 cost= 0.335739001
Epoch: 025 cost= 0.333702818
Optimization Finished!
Accuracy: 0.9146
To view summaries in the Tensorboard, run the command line:
-->tensorboard --logdir=/tmp/tensorflow_tensorboard
Then open http://0.0.0.0:6006/ into your web browser
```

要在 TensorBoard 中查看统计数据,需要在终端中使用以下命令。

```
tensorboard --logdir=/tmp/tensorflow_tensorboard
```

然后,在 Web 浏览器中打开 http://0.0.0.0:6006/。

当打开 TensorBoard 时,你应该得到图 4.8 所示的界图。

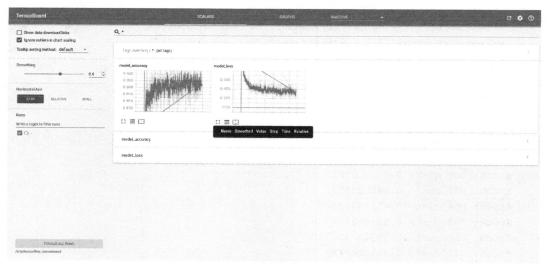

图 4.8　打开的 TensorBoard 界面

这显示了我们正在监控的变量,例如模型的准确性、模型的拟合度如何变得更高、模型的损失值以及模型在整个训练过程中的变化情况。通过这个例子发现,这是一个正常的学习过程。但有时我们会发现准确度和模型损失值或者想要跟踪一些其他变量在整个会话期间会随机发生变化以及它们如何变化,所以 TensorBoard 有助于我们发现任何随机性变化或者其他错误。

另外，如果切换到 TensorBoard 中的 GRAPHS 选项卡，将会看到在前面的代码中构建的计算图，如图 4.9 所示。

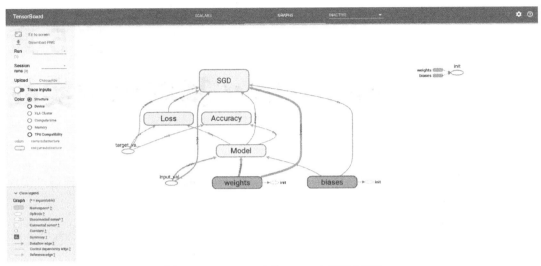

图 4.9　TensorBoard 中的 GRAPHS 选项卡界面

4.7　总结

本章介绍了在 Ubuntu 和 Mac 系统上 TensorFlow 的安装过程，概述了 TensorFlow 编程模型，并解释了可用于构建复杂模型的不同类型的基本节点，以及如何使用会话对象从 TensorFlow 获取输出。另外，本章还介绍了 TensorBoard 以及为什么它有助于调试和分析复杂的深度学习应用程序。

第 5 章将对神经网络和多层神经网络背后的机制做一个详细的介绍，同时还将介绍一些 TensorFlow 的基本示例，并演示如何将它用于回归和分类问题。

第 5 章
TensorFlow 基础示例实战

本章将解释 TensorFlow 背后的主要计算概念,即计算图模型,并演示如何通过实现线性回归和逻辑回归使读者走上学习深度学习知识的正轨。

本章将介绍以下内容。
- 神经元的结构。
- 激活函数。
- 前馈神经网络。
- 需要多层网络的原因。
- TensorFlow 术语回顾。
- 构建与训练线性回归模型。
- 构建与训练逻辑回归模型。

本章将首先解释单个神经元实际上可以做什么,并基于此,引出工程对多层神经网络的需求。接下来,本章将更详细地介绍 TensorFlow 中使用/可用的主要概念和工具,以及如何使用这些工具构建简单的示例,如线性回归模型和逻辑回归模型。

5.1 神经元的结构

神经网络是一种计算模型,它主要受到人类大脑的神经网络处理传入信息的方式的启发。神经网络在机器学习研究(特别是深度学习)和行业应用方面取得了巨大突破,例如在计算机视觉、语音识别和文本处理方面取得的突破性成果。在本章中,我们将尝试研究一种特定类型的神经网络，它称为**多层感知机**。

生物学中的激活和连接

人类大脑的基本计算单元称为**神经元**，神经系统中有大约 860 亿个神经元，它们通过 $10^{14} \sim 10^{15}$ 个突触相连。

图 5.1 所示为一种生物神经元。图 5.2 所示为相应的数学模型。在生物神经元中，每个神经元接收来自其树突的输入信号，然后沿其轴突产生输出信号，其中轴突最后被分离并通过突触连接到其他神经元。

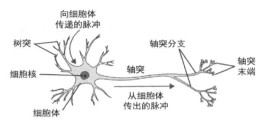

图 5.1　大脑的计算单元（图片源自 GitHub 网站）

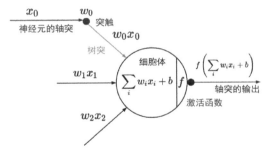

图 5.2　大脑计算单元的数学模型（图片源自 GitHub 网站）

在神经元的相应数学计算模型中，沿着轴突传输的信号 x_0 与其他神经元树突上的突触强度 ω_0 经过乘法运算得到 $x_0\omega_0$。这个想法的目的是使突触的权重/强度 w 通过网络进行学习，它们也是控制特定神经元对另一个神经元影响的重要因素。

此外，在图 5.2 的基本计算模型中，树突将信号传送到主细胞体，然后在那里将它们全部相加。如果最终结果高于某个阈值，则可以在计算模型中激活神经元。

此外，值得一提的是，因为需要控制轴突上输出信号的频率，所以需要使用**激活函数**。实际上，激活函数的一个共同选择是 sigmoid 函数 σ，因为它接受实值输入（累加之后的信号强度）并将输入压缩到 0～1。读者将在稍后看到这些激活函数的细节。

生物学模型所对应的数学模型如图 5.2 所示。

神经网络中的基本计算单元是神经元,也称为**节点**或**单元**。它从一些其他节点或外部源接收输入并计算输出。每个输入都有一个相关的**权重**(w),权重是根据其相对于其他输入的重要性来分配的。节点将使用函数 f(稍后定义)应用于其输入的加权和。

综上所述,神经网络的计算单元通常称作**神经元、节点**或**单元**。

每个神经元是从先前的神经元或者外部源接收输入,然后它对该输入进行一些处理以产生所谓的激活状态。该神经元的每个输入都与其对应的权重 w 相关联,权重 w 表示连接的强度,也表示该输入的重要性。

因此,神经网络中神经元的最终输出是输入及其对应权重 w 的加权和,然后神经元通过激活函数传递加权求和后的输出,如图 5.3 所示。

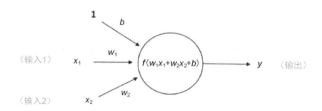

神经元的输出=$y=f(w_1x_1+w_2x_2+b)$

图 5.3　单个神经元

5.2　激活函数

神经元的输出如图 5.3 所示。通过激活函数,该神经元输出非线性结果。该函数 f 就称为**激活函数**。激活函数的主要作用如下。

- 把神经元的输出变为非线性表示方式。这很重要,因为现实世界中的大多数数据都是非线性的,我们肯定想要神经元学习这些非线性表示。
- 将输出压缩到特定范围内。

每个激活函数(或非线性表示方式)都会对输入的单个数执行某个固定的数学运算。我们可能会在实践中遇到多种激活函数。

因此，这里将简要介绍最常见的激活函数。

5.2.1 sigmoid

在历史上，sigmoid 激活函数在研究人员中广泛使用。此函数接受实值输入并将它压缩到[0,1]，如图 5.4 所示。

$$\sigma(x) = \frac{1}{1+e^{-x}}$$

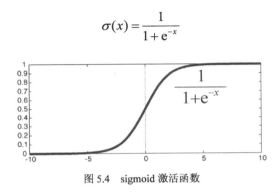

图 5.4 sigmoid 激活函数

5.2.2 tanh

tanh 是另一种能够接受一些负值输入的激活函数。tanh 接受一个实值输入并将它们压缩到[-1,1]，如图 5.5 所示。

$$\tanh(x) = 2\sigma(2x) - 1$$

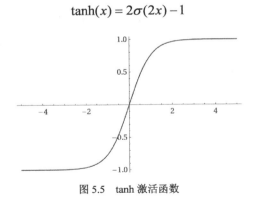

图 5.5 tanh 激活函数

5.2.3 ReLU

修正线性单元（Rectified Linear Unit，ReLU）不输出负值，它接受实值输入并将其阈值设置为零（将负值替换为零），如图 5.6 所示。

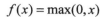

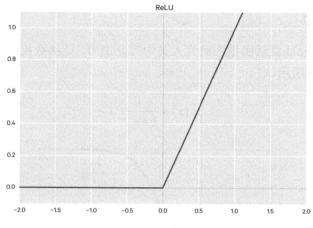

图 5.6　ReLU 激活函数

下面讨论**偏差**（bias）的重要性。偏差的主要功能是为每个节点提供一个可训练的常数值（除了节点接受的正常输入之外）。请参考 stackoverflow 网站，了解偏差在神经元中的作用以及更多信息。

5.3　前馈神经网络

前馈神经网络（feed-forward neural network）是第一种最简单的人工神经网络。它包含多层神经元（节点）。相邻层的节点之间有连接或边。所有这些连接都具有与之相关的权重。

图 5.7 所示为一个前馈神经网络的简单例子。

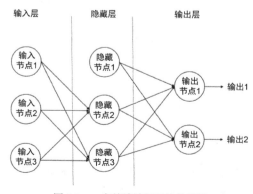

图 5.7　一个前馈神经网络的例子

在前馈网络中，信息仅在一个方向上传播，从输入节点向前移动，通过隐藏节点（如果有的话）移动到输出节点。网络中没有环或循环（前馈网络的这种属性不同于循环神经网络，循环神经网络中节点之间的连接形成一个循环）。

5.4 需要多层网络的原因

多层感知器（multi-layer perceptron，MLP）包含一个或多个隐藏层（除了一个输入层和一个输出层之外）。和单层感知器只能学习线性函数不同，MLP 也可以学习非线性函数。

图 5.8 所示为具有单个隐藏层的 MLP。注意，所有连接都具有与之关联的权重，但图中仅显示了 3 个权重（w_0、w_1 和 w_2）。

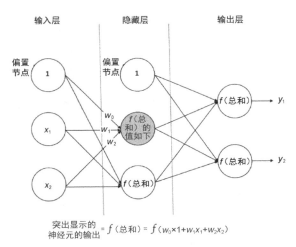

突出显示的神经元的输出 $= f(总和) = f(w_0 \times 1 + w_1 x_1 + w_2 x_2)$

图 5.8 含有单个隐藏层的多层感知器

输入层有 3 个节点，偏差节点的值为 1，另外两个节点以 x_1 和 x_2 作为外部输入（这取决于输入数据集的数值）。如前所述，输入层中不执行任何计算，因此输入层中节点的输出分别为 1、x_1 和 x_2，把它们 3 个输入**隐藏层**。

隐藏层也有 3 个节点，偏差节点的输出为 1。隐藏层中其他两个节点的输出取决于输入层（1、x_1 和 x_2）的输出以及连接线（边）上相关的权重。注意，f 指的是激活函数。然后将这些输出传到输出层中的节点。

输出层有两个节点，它们从隐藏层获取输入，并执行与突出显示的隐藏节点相似的计算。得到的这些计算结果（y_1 和 y_2）可充当多层感知器的输出。

给定一组特征 $X = (x1, x2, \cdots)$ 和目标 y，多层感知器可以学习特征与目标之间的关系，并用于分类或回归。

这里举个例子以更好地理解多层感知器，假设关于某学生成绩的数据集如表 5.1 所示。

表 5.1 关于学生成绩的数据集

学习的小时数	期中考试分数	期末考试结果
35	67	通过
12	75	未通过
16	89	通过
45	56	通过
10	90	未通过

左侧的两个输入列显示学生学习的小时数和学生的期中考试分数。右侧的输出列可以有两个值——1 或 0，表示学生是否通过期末考试。例如，我们可以看到，如果学生学习了 35 小时并且在期中考试获得了 67 分，那他最终会通过期末考试。

现在，假设读者想要预测一名学习 25 小时并且期中考试有 70 分的学生是否会通过期末考试，该样本如表 5.2 所示。

表 5.2 学生期末考试结果未知的样本

学习的小时数	期中考试分数	期末考试结果
26	70	?

这是一个二元分类问题，使用 MLP 可以从给定的示例（训练数据）中学习并在给定新数据点的情况下做出正确的预测。我们很快就会看到 MLP 如何学习这种关系。

5.4.1 训练 MLP——反向传播算法

多层感知器学习的过程称为**反向传播**算法（backpropagation algorithm）。这里建议阅读 Quora 网站上 Hemanth Kumar 在 "How do you explain back propagation algorithm to a beginner inneural network" 文章中给出的答案，它清楚地解释了反向传播算法。

误差的反向传播（通常缩写为 BackProp）是训练人工神经网络（ANN）的几种方式之一。它是一种监督的训练方法，这意味着它从有标记的训练数据中学习（有一个监督者来指导学习）。

简单来说，BackProp 就像"从差错中学习"。每当出错时，监督者都会纠正 ANN。

ANN 由不同层（输入层、中间隐藏层和输出层）中的节点组成。相邻层的节点之间的连接具有与它们相关联的权重。学习的目标是为这些边分配正确的权重。给定输入向量，这些权重确定输出向量是什么。在监督学习中，标记训练集。这意味着对于某些给定的输入，我们知道期望/预期的输出（标签）。

BackProp 算法：

最初，所有连接权重都是随机分配的。对于训练数据集中的每个输入，激活 ANN 并观察其输出。将此输出与我们已知的期望输出进行比较，并将误差"传播"回上一层。ANN 记下该误差并相应地"调整"权重。重复该过程直到输出的误差低于预定阈值。

一旦上述算法终止，我们就会有一个"学习好"的 ANN，我们认为它已准备好接受"新"输入。可以认为这个 ANN 从几个样本（标记数据）及其误差（误差传播）中学到了东西。

——Hemanth Kumar

现在我们已经了解了反向传播算法的工作原理，回到之前关于学生成绩的数据集。

图 5.8 所示的 MLP 在输入层中具有两个节点，它们表示输入小时数和期中考试成绩。它还有一个带有两个节点的隐藏层。输出层也有两个节点，上侧节点输出期末考试通过（pass）的概率，而下侧节点输出未通过（fail）的概率。

分类应用会广泛使用 softmax 函数（参见 GitHub 网站）作为 MLP 输出层中的激活函数，以确保输出的概率总和为 1。softmax 函数接受任意实值的向量输入，并将它压缩成 0~1 的值向量。因此，在这种情况下：

$$通过的概率 + 未通过的概率 = 1$$

5.4.2 前馈传播

神经网络中的所有权重都是随机初始化的。这里考虑一个特定的隐藏层节点并将它称为 V。假设从输入层到该节点的连接的权重分别是 w_1、w_2 和 w_3（见图 5.8）。

然后神经网络将第一个训练样本作为输入（对于输入 35 和 67，通过期末考试的概率是 1）。

- 神经网络的输入是[35,67]。
- 神经网络所需的输出（目标）是[1,0]。

所关注的节点的输出 V 可以按照以下方式计算（f 是诸如 sigmoid 的激活函数）。

$$V = f(1w_1 + 35w_2 + 67w_3)$$

类似地，还可以同时计算隐藏层中另一节点的输出。隐藏层中两个节点的输出充当输出层中两个节点的输入。这使我们能够从输出层中的两个节点计算输出概率。

假设输出层中两个节点的输出概率分别为 0.4 和 0.6（因为权重是随机分配的，所以输出也是随机的）。可以看到计算出的概率（0.4 和 0.6）与期望的概率（分别为 1 和 0）相差甚远，因此，可以认为这次神经网络的输出不正确。

5.4.3 反向传播和权值更新

这里需要计算输出节点处的总误差，并使用反向传播算法通过神经网络将这些误差的梯度反向传播回去。然后，使用诸如梯度下降的优化方法来调整网络中的所有权重，目的是减少输出层的误差。

假设与所考虑的节点相关联的新权重是 w_4、w_5 和 w_6（在反向传播和调整权重之后）。

如果现在将相同的样本作为输入传到神经网络，则应该比初始运行得更好，因为现在已经优化权重以最小化预测中的误差了。与先前的[0.6，−0.4]相比，输出节点处的误差现在减少到[0.2，−0.2]。这意味着网络已经学会正确分类第一个训练样本。

此后在数据集中使用其他训练样本重复此过程，这时可以认为网络已经学会了这些例子。

如果我们现在想要预测一个学习 25 小时并且在期中考试中得 70 分的学生是否会通过期末考试，可以使用前向传播步骤来得到通过和未通过的概率。

这里不使用数学方程式，也不解释诸如梯度下降等概念，而是试图解释算法的思想。

5.5 TensorFlow 术语回顾

本节将概述 TensorFlow 库以及基本的 TensorFlow 应用程序的结构。TensorFlow 是一个用于创建大规模机器学习应用程序的开源库，它可以模拟各种硬件（从 Android 设备到异构多 GPU 系统）上的计算。

TensorFlow 使用了特殊结构，以便在不同的设备（如 CPU 和 GPU）上执行代码。把计算定义为计算图，每个计算图由运算（也称为计算）组成，因此每当我们使用 TensorFlow 时，都会在计算图中定义一系列运算。

要运行这些运算，我们需要将计算图部署到会话中。会话会转化运算并将它们传递给设备以便执行。

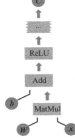

例如，图 5.9 所示为 TensorFlow 中的计算图。W、x 和 b 是该图中边上的张量。MatMul 表示对张量 W 和 x 的运算，之后调用 Add，这里用 b 添加前一个运算符的结果。把每个运算的结果张量传给下一个运算，直到运算结束，在结束处可以获得所需的结果。

图 5.9 一个简单的 Tensorflow 计算图

为了使用 TensorFlow，我们需要导入库。这里将它命名为 tf，以便可以通过写入 tf. 和模块的名称来访问模块。

```
import tensorflow as tf
```

第一个图的创建将从使用源操作开始，源操作不需要任何输入。这些源操作将其信息传递给其他操作，这些操作实际上将运行计算。

以下是创建可以输出数字的两个源操作的代码，这两个源操作为 A 和 B。

```
A = tf.constant([2])
B = tf.constant([3])
```

之后，还将定义一个简单的计算操作 tf.add()，用于对两个元素求和。也可以使用 C = A + B，如以下代码所示。

```
C = tf.add(A,B)
#C = A + B is also a way to define the sum of the terms
```

由于计算图需要在会话中执行，因此需要创建一个会话对象。

```
session = tf.Session()
```

要观察计算图，可以运行会话来获得先前定义的 C 操作的结果。

```
result = session.run(C)
print(result)
```

输出如下。

[5]

读者可能认为将两个数字相加需要做很多工作，但了解 TensorFlow 的基本结构非常重要，这有助于定义所需的任何计算。TensorFlow 的结构允许它处理不同设备（CPU 或 GPU）上的计算，甚至集群中的计算。如果要了解更多相关信息，可以运行方法 **tf.device()**。

同时我们也可以随意尝试使用 TensorFlow 的结构，以便更好地了解它的工作原理。如果需要了解 TensorFlow 支持的所有数学运算的列表，可以查看文档。

到目前为止，我们了解了 TensorFlow 的结构以及如何创建基本应用程序。

5.5.1 使用 Tensorflow 定义多维数组

接下来可以使用如下方法定义数组。

```
salar_var = tf.constant([4])
vector_var = tf.constant([5,4,2])
matrix_var = tf.constant([[1,2,3],[2,2,4],[3,5,5]])
tensor = tf.constant( [ [[1,2,3],[2,3,4],[3,4,5]] ,
[[4,5,6],[5,6,7],[6,7,8]] , [[7,8,9],[8,9,10],[9,10,11]] ] )
with tf.Session() as session:
    result = session.run(salar_var)
    print "Scalar (1 entry):\n %s \n" % result
    result = session.run(vector_var)
    print "Vector (3 entries) :\n %s \n" % result
    result = session.run(matrix_var)
    print "Matrix (3x3 entries):\n %s \n" % result
    result = session.run(tensor)
    print "Tensor (3x3x3 entries) :\n %s \n" % result
```

输出如下。

```
Scalar (1 entry):
 [2]

Vector (3 entries) :
 [5 6 2]

Matrix (3x3 entries):
 [[1 2 3]
```

```
  [2 3 4]
  [3 4 5]]

Tensor (3x3x3 entries) :
[[[ 1  2  3]
  [ 2  3  4]
  [ 3  4  5]]

 [[ 4  5  6]
  [ 5  6  7]
  [ 6  7  8]]

 [[ 7  8  9]
  [ 8  9 10]
  [ 9 10 11]]]
```

既然我们已经了解了这些数据结构,这里建议根据它们的结构类型使用它们实现一些以前的函数来查看其行为。

```
Matrix_one = tf.constant([[1,2,3],[2,3,4],[3,4,5]])
Matrix_two = tf.constant([[2,2,2],[2,2,2],[2,2,2]])
first_operation = tf.add(Matrix_one, Matrix_two)
second_operation = Matrix_one + Matrix_two
with tf.Session() as session:
    result = session.run(first_operation)
    print "Defined using tensorflow function :"
    print(result)
    result = session.run(second_operation)
    print "Defined using normal expressions :"
    print(result)
```

输出如下。

```
Defined using tensorflow function :
[[3 4 5]
 [4 5 6]
 [5 6 7]]

Defined using normal expressions :
[[3 4 5]
 [4 5 6]
 [5 6 7]]
```

通过常规符号定义以及 tensorflow 函数,我们能够实现对应元素相乘,它也称为 **Hadamard 乘积**。但是,如果想要得到常规矩阵的乘积呢?这里需要使用另一个名为 tf.matmul() 的 TensorFlow 函数。

```
Matrix_one = tf.constant([[2,3],[3,4]])
Matrix_two = tf.constant([[2,3],[3,4]])
first_operation = tf.matmul(Matrix_one, Matrix_two)
with tf.Session() as session:
    result = session.run(first_operation)
    print "Defined using tensorflow function :"
    print(result)
```

输出如下。

```
Defined using tensorflow function :
[[13 18]
 [18 25]]
```

我们也可以自己定义这个乘法,但是有一个函数已经做到了,所以不需要重新发明"轮子"。

5.5.2 为什么使用张量

张量结构允许我们按照自己的方式自由地塑造数据集。

图像中信息的编码方式在处理图像时非常有用。

说到图像,很容易认识到它具有高度和宽度,因此用二维结构(矩阵)表示包含在其中的信息是有意义的。考虑到图像有颜色,要添加有关颜色的信息,就需要另一个维度,那时张量会变得特别有用。

把图像编码为颜色通道,图像数据在给定点的颜色通道中以每种颜色的强度表示,最常见的是 RGB(表示红色、绿色和蓝色)。图像中包含的信息是图像宽度和高度中每种通道颜色的强度,原始图像及其 3 个颜色通道如图 5.10(a)~(d)所示。

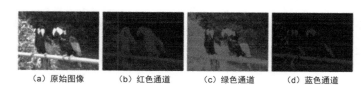

(a)原始图像　　(b)红色通道　　(c)绿色通道　　(d)蓝色通道

图 5.10　一张图像中不同的颜色通道

因此，在每个点宽度和高度对应的红色通道强度可以用矩阵表示，对于蓝色和绿色通道也是如此。我们最终得到 3 个矩阵，当它们组合在一起时，便形成了一个张量。

5.5.3 变量

现在我们对数据结构已经很熟悉了，接下来看看 TensorFlow 如何处理变量。

要定义变量，这里使用命令 tf.variable()。为了能够在计算图中使用变量，有必要在会话中运行计算图之前对它们进行初始化。这是通过运行 tf.global_variables_initializer() 来完成的。

要更新变量的值，只须执行一个赋值操作。

```
state = tf.Variable(0)
```

首先创建一个简单的计数器，一个一次增加一个单位的变量。

```
one = tf.constant(1)
new_value = tf.add(state, one)
update = tf.assign(state, new_value)
```

必须通过在启动图形后运行初始化操作来初始化变量。首先要将初始化操作添加到图中。

```
init_op = tf.global_variables_initializer()
```

然后启动一个运行图的会话。

这里首先初始化变量，然后输出状态变量的初始值，最后运行更新状态变量的操作，并在每次更新后输出结果。

```
with tf.Session() as session:
  session.run(init_op)
  print(session.run(state))
  for _ in range(3):
     session.run(update)
     print(session.run(state))
```

输出如下。

```
0
1
```

2
3

5.5.4 占位符

现在,我们已经知道如何在 TensorFlow 中操作变量,但是如何在 TensorFlow 模型之外提供数据呢?

如果要从模型外部向 TensorFlow 模型提供数据,则需要使用占位符。

那么,这些占位符是什么?它们做了什么?占位符可以看作模型中的孔,我们可以传递数据到孔中。使用 tf.placeholder(datatype) 创建占位符,其中 datatype 指定数据类型(整数、浮点数、字符串和布尔值)及其精度(8 位、16 位、32 位和 64 位)。

可以使用相应 Python 语法定义每种数据类型,如表 5.3 所示。

表 5.3 Tensorflow 中不同数据类型的定义

数据类型	Python 类型	描述
DT_FLOAT	tf.float32	32 位浮点数
DT_DOUBLE	tf.float64	64 位浮点数
DT_INT8	tf.int8	8 位有符号整数
DT_INT16	tf.int16	16 位有符号整数
DT_INT32	tf.int32	32 位有符号整数
DT_INT64	tf.int64	64 位有符号整数
DT_UINT8	tf.uint8	8 位无符号整数
DT_STRING	tf.string	可变长度的字节数组,每一个张量元素都是一个字节数组
DT_BOOL	tf.bool	布尔型
DT_COMPLEX64	tf.complex64	由两个 32 位浮点数(实部和虚部)组成的复数
DT_COMPLEX128	tf.complex128	由两个 64 位浮点数(实部和虚部)组成的复数
DT_QINT8	tf.qint8	用于量化 Ops 的 8 位有符号整数
DT_QINT32	tf.qint32	用于量化 Ops 的 32 位有符号整数
DT_QUINT8	tf.quint8	用于量化 Ops 的 8 位无符号整数

接下来创建一个占位符。

```
a=tf.placeholder(tf.float32)
```

同时也定义一个简单的乘法操作。

b=a*2

现在，需要定义并运行会话。因为我们在初始化会话时在模型中创建了一个用于传递数据的洞，所以这里不得不对数据进行填充，否则会出现错误。

为了将数据传递给模型，需要使用另一个参数 feed_dict 来调用会话。我们应该在其中传递一个包含每个占位符名称及其相应数据的字典。

```
with tf.Session() as sess:
    result = sess.run(b,feed_dict={a:3.5})
    print result
```

输出如下。

7.0

因为 TensorFlow 中的数据以多维数组的形式传递，所以这里可以通过占位符传递任何类型的张量来获得简单乘法运算的结果。

```
dictionary={a: [ [ [1,2,3],[4,5,6],[7,8,9],[10,11,12] ] , [
[13,14,15],[16,17,18],[19,20,21],[22,23,24] ] ] }
with tf.Session() as sess:
    result = sess.run(b,feed_dict=dictionary)
    print result
```

输出如下。

```
[[[  2.   4.   6.]
  [  8.  10.  12.]
  [ 14.  16.  18.]
  [ 20.  22.  24.]]

 [[ 26.  28.  30.]
  [ 32.  34.  36.]
  [ 38.  40.  42.]
  [ 44.  46.  48.]]]
```

5.5.5 操作

操作是表示计算图上张量之间数学运算的节点。这些操作可以是任何类型的函数，如张量加法和张量减法，或者可能是激活函数。

tf.matmul、tf.add 和 tf.nn.sigmoid 是 TensorFlow 中的一些操作。它们类似于 Python

中的函数，但直接作用于张量，每个函数都执行特定的操作。

其他操作可以在 TensorFlow 官网上轻松找到。

下面来看看其中的一些操作。

```
a = tf.constant([5])
b = tf.constant([2])
c = tf.add(a,b)
d = tf.subtract(a,b)
with tf.Session() as session:
    result = session.run(c)
    print 'c =: %s' % result
    result = session.run(d)
    print 'd =: %s' % result
```

输出如下。

```
c =: [7]
d =: [3]
```

tf.nn.sigmoid 是一个激活函数，它有点复杂，但是这个函数有助于学习模型来评估信息的好坏。

5.6 构建与训练线性回归模型

根据第 2 章对线性回归的解释，这里将依赖该定义来构建简单的线性回归模型。

现在从导入此次实现必需的包开始。

```
import numpy as np
import tensorflow as tf
import matplotlib.patches as mpatches
import matplotlib.pyplot as plt
plt.rcParams['figure.figsize'] = (10, 6)
```

接下来定义自变量。

```
input_values = np.arange(0.0, 5.0, 0.1)
input_values

Output:
 array([ 0. ,  0.1, 0.2, 0.3, 0.4, 0.5, 0.6, 0.7, 0.8, 0.9, 1. ,
```

```
            1.1, 1.2, 1.3, 1.4, 1.5, 1.6, 1.7, 1.8, 1.9, 2. , 2.1,
            2.2, 2.3, 2.4, 2.5, 2.6, 2.7, 2.8, 2.9, 3. , 3.1, 3.2,
            3.3, 3.4, 3.5, 3.6, 3.7, 3.8, 3.9, 4. , 4.1, 4.2, 4.3,
            4.4, 4.5, 4.6, 4.7, 4.8, 4.9])

##You can adjust the slope and intercept to verify the changes in the graph
weight=1
bias=0
output = weight*input_values + bias
plt.plot(input_values,output)
plt.ylabel('Dependent Variable')
plt.xlabel('Indepdendent Variable')
plt.show()
```

输出如下。

输出如图 5.11 所示。

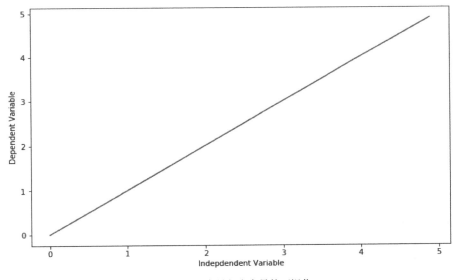

图 5.11　因变量与自变量的可视化

现在，看看如何将线性回归模型解释为 TensorFlow 代码。

使用 Tensorflow 实现线性回归

对于第一部分，本节将生成随机数据点并定义线性关系。我们将使用 TensorFlow 来调整并获得正确的参数。

```
input_values = np.random.rand(100).astype(np.float32)
```

这个例子使用的公式是:

$$Y = 2X + 3$$

这个公式没什么特别之处,它只是用来生成数据点的模型。实际上,我们可以将参数更改为想要的任何内容,稍后将执行此操作。这里将会在点上添加一些高斯噪声,使它更有趣。

```
output_values = input_values * 2 + 3
output_values = np.vectorize(lambda y: y + np.random.normal(loc=0.0,
scale=0.1))(output_values)
```

下面列出一些生成的数据。

```
list(zip(input_values,output_values))[5:10]
```

输出如下。

```
[(0.25240293, 3.474361759429548),
(0.946697, 4.980617375175061),
(0.37582186, 3.650345806087635),
(0.64025956, 4.271037640404975),
(0.62555283, 4.37001850440196)]
```

首先,需要随机地对变量 weight 和 bias 进行初始化,并定义线性函数。

```
weight = tf.Variable(1.0)
bias = tf.Variable(0.2)
predicted_vals = weight * input_values + bias
```

在典型的线性回归模型中,需要最小化可训练的线性函数的预测值减去目标值(已拥有的数据)的平方误差,因此这里使用该线性函数的预测值来计算最小损失。

为了计算损失值,这里使用 tf.reduce_mean()。此函数求多维张量的平均值,结果可以具有不同的维度。

```
model_loss = tf.reduce_mean(tf.square(predicted_vals - output_values))
```

5.6 构建与训练线性回归模型

然后,定义优化器方法。这里将使用简单的梯度下降法,学习率为 0.5。

现在定义计算图的训练方法,但我们将使用什么方法来最小化损失呢?它就是 tf.train.GradientDescentOptimizer。

.minimize()函数将最小化优化器的误差函数,从而产生更好的模型。

```
model_optimizer = tf.train.GradientDescentOptimizer(0.5)
train = model_optimizer.minimize(model_loss)
```

注意,别忘记在运行计算图之前初始化变量。

```
init = tf.global_variables_initializer()
sess = tf.Session()
sess.run(init)
```

下面正式开始优化并运行计算图。

```
train_data = []
for step in range(100):
    evals = sess.run([train,weight,bias])[1:]
    if step % 5 == 0:
        print(step, evals)
        train_data.append(evals)
```

输出如下。

```
(0, [2.5176678, 2.9857566])
(5, [2.4192538, 2.3015416])
(10, [2.5731843, 2.221911])
(15, [2.6890132, 2.1613526])
(20, [2.7763696, 2.1156814])
(25, [2.8422525, 2.0812368])
(30, [2.8919399, 2.0552595])
(35, [2.9294133, 2.0356679])
(40, [2.957675, 2.0208921])
(45, [2.9789894, 2.0097487])
(50, [2.9950645, 2.0013444])
(55, [3.0071881, 1.995006])
(60, [3.0163314, 1.9902257])
(65, [3.0232272, 1.9866205])
(70, [3.0284278, 1.9839015])
(75, [3.0323503, 1.9818509])
(80, [3.0353084, 1.9803041])
```

```
(85, [3.0375392, 1.9791379])
(90, [3.039222, 1.9782581])
(95, [3.0404909, 1.9775947])
```

下面对拟合数据点的训练过程进行可视化。

```
print('Plotting the data points with their corresponding fitted line...')
converter = plt.colors
cr, cg, cb = (1.0, 1.0, 0.0)

for f in train_data:

    cb += 1.0 / len(train_data)
    cg -= 1.0 / len(train_data)
    if cb > 1.0: cb = 1.0

    if cg < 0.0: cg = 0.0

    [a, b] = f
    f_y = np.vectorize(lambda x: a*x + b)(input_values)
    line = plt.plot(input_values, f_y)
    plt.setp(line, color=(cr,cg,cb))

plt.plot(input_values, output_values, 'ro')
green_line = mpatches.Patch(color='red', label='Data Points')
plt.legend(handles=[green_line])
plt.show()
```

输出的可视化结果如图 5.12 所示。

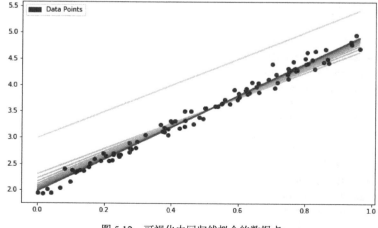

图 5.12　可视化由回归线拟合的数据点

5.7 构建与训练逻辑回归模型

同样基于第 2 章中对逻辑回归的解释，这里将在 TensorFlow 中实现逻辑回归算法。简单地说，逻辑回归通过 logistic/sigmoid 传递输入，随后将结果视为概率，如图 5.13 所示。

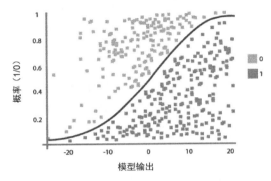

图 5.13　区分两个线性可分的类，0 和 1

在 TensorFlow 中使用逻辑回归

为了在 TensorFlow 中使用逻辑回归，首先需要导入将要使用的所有库：

```
import tensorflow as tf

import pandas as pd

import numpy as np
import time
from sklearn.datasets import load_iris
from sklearn.cross_validation import train_test_split
import matplotlib.pyplot as plt
```

接下来，需要加载将要使用的数据集。在这种情况下，我们使用内置的 iris 数据集。因此，没有必要进行任何预处理，可以直接进行操作。这里将数据集分为 x 和 y，然后随机分为训练 x 和 y，测试 x 与 y：

```
iris_dataset = load_iris()
iris_input_values, iris_output_values = iris_dataset.data[:-1,:],
```

```
iris_dataset.target[:-1]
iris_output_values= pd.get_dummies(iris_output_values).values
train_input_values, test_input_values, train_target_values,
test_target_values = train_test_split(iris_input_values,
iris_output_values, test_size=0.33, random_state=42)
```

现在，需要定义 x 和 y，这些占位符将保存 iris 数据（特征和标签矩阵），并将它们传递到算法的不同部分。可以将占位符视为需要插入数据的空壳，同时还需要为它们提供与数据形状相对应的形状。稍后，将通过 feed_dict（feed 字典）向占位符提供数据，并将数据插入这些占位符中。

1. 为什么使用占位符

TensorFlow 的这一特性使我们能够创建一种算法来接受数据并了解数据的形状，而无须知道插入的数据量。当在训练中插入批量数据时，在不改变整个算法的情况下，我们可以在一步中轻松调整训练的样本数量。

```
# numFeatures is the number of features in our input data.
# In the iris dataset, this number is '4'.
num_explanatory_features = train_input_values.shape[1]

# numLabels is the number of classes our data points can be in.
# In the iris dataset, this number is '3'.
num_target_values = train_target_values.shape[1]

# Placeholders
# 'None' means TensorFlow shouldn't expect a fixed number in that dimension
input_values = tf.placeholder(tf.float32, [None, num_explanatory_features])
# Iris has 4 features, so X is a tensor to hold our data.
output_values = tf.placeholder(tf.float32, [None, num_target_values])
# This will be our correct answers matrix for 3 classes.
```

2. 设置模型的权重和偏差

与线性回归非常相似，对于逻辑回归我们需要一个共享变量权重矩阵。这里将 W 和 b 初始化为零张量。由于要学习 W 和 b，因此它们的初始值并不重要。这些变量是定义回归模型结构的对象，可以在训练变量后保存它们，以便以后可以重用它们。

这里将两个 TensorFlow 变量定义为参数，这些变量将保持逻辑回归的权重和偏差，并且它们将在训练期间不断更新。

请注意，W 的形状为[4,3]，这是因为希望 4 维输入向量乘以它，以产生不同类别的三维输出向量。b 的形状为[3]，因此可以将它添加到输出中。此外，与占位符（基本上是等待输入数据的空壳）不同，TensorFlow 变量需要使用值（如零）进行初始化。

```
#Randomly sample from a normal distribution with standard deviation .01

weights =
tf.Variable(tf.random_normal([num_explanatory_features,num_target_values],
                              mean=0,
                              stddev=0.01,
                              name="weights"))

biases = tf.Variable(tf.random_normal([1,num_target_values],
                              mean=0,
                              stddev=0.01,
                              name="biases"))
```

3. 逻辑回归模型

现在定义一些操作以便正确地运行逻辑回归。通常认为逻辑回归是单个等式。

$$\hat{y} = \text{sigmoid}(WX + b)$$

然而，为了清楚起见，它可以分为 3 个主要组成部分。

- 权重乘以特征矩阵的乘法操作。
- 加权特征和偏差项的总和。
- 应用的 sigmoid 函数。

我们会发现这些组件被定义为 3 个独立的操作。

```
# Three-component breakdown of the Logistic Regression equation.
# Note that these feed into each other.
apply_weights = tf.matmul(input_values, weights, name="apply_weights")
add_bias = tf.add(apply_weights, biases, name="add_bias")
activation_output = tf.nn.sigmoid(add_bias, name="activation")
```

正如之前看到的，这里将要使用的函数是逻辑函数，它将应用了权重和偏差的数据作为自己的输入数据。在 TensorFlow 中，此函数实现为 nn.sigmoid 函数。实际上，nn.sigmoid 函数将加权输入与偏差拟合为 0%～100%的曲线，该曲线是我们想要的概率函数。

4. 训练

学习算法的目标是搜索最佳权重向量（w）。此搜索是一个优化问题，目的是寻找可以优化误差/成本函数的方法。

在实际应用中，模型的成本或损失函数可能会告诉我们模型非常糟糕，需要最小化这个函数。我们可以遵循不同的损失或成本标准。此实现将使用**均方误差**（mean squared error，MSE）作为损失函数。

为了完成最小化损失函数的任务，我们将使用梯度下降算法。

5. 成本函数

在定义成本函数（cost function）之前，需要定义要训练的时间以及学习率。

```
#Number of training epochs
num_epochs = 700
# Defining our learning rate iterations (decay)
learning_rate = tf.train.exponential_decay(learning_rate=0.0008,
                                            global_step=1,
decay_steps=train_input_values.shape[0],

                                            decay_rate=0.95,
                                            staircase=True)

# Defining our cost function - Squared Mean Error
model_cost = tf.nn.l2_loss(activation_output - output_values,
name="squared_error_cost")
# Defining our Gradient Descent
model_train =
tf.train.GradientDescentOptimizer(learning_rate).minimize(model_cost)
```

现在，通过会话变量执行计算图。

首先，需要使用 tf.initialize_all_variables()用零或随机值初始化权重和偏差项。这个初始化步骤将成为计算图中的一个节点。当将计算图放入一个会话时，将执行该操作并创建变量。

```
# tensorflow session
sess = tf.Session()
```

```
# Initialize our variables.
init = tf.global_variables_initializer()
sess.run(init)

#We also want some additional operations to keep track of our model's
efficiency over time. We can do this like so:
# argmax(activation_output, 1) returns the label with the most probability
# argmax(output_values, 1) is the correct label
correct_predictions =
tf.equal(tf.argmax(activation_output,1),tf.argmax(output_values,1))

# If every false prediction is 0 and every true prediction is 1, the
average returns us the accuracy
model_accuracy = tf.reduce_mean(tf.cast(correct_predictions, "float"))

# Summary op for regression output
activation_summary = tf.summary.histogram("output", activation_output)

# Summary op for accuracy
accuracy_summary = tf.summary.scalar("accuracy", model_accuracy)

# Summary op for cost
cost_summary = tf.summary.scalar("cost", model_cost)

# Summary ops to check how variables weights and biases are updating after
each iteration to be visualized in TensorBoard
weight_summary = tf.summary.histogram("weights",
weights.eval(session=sess))
bias_summary = tf.summary.histogram("biases", biases.eval(session=sess))

merged = tf.summary.merge([activation_summary, accuracy_summary,
cost_summary, weight_summary, bias_summary])
writer = tf.summary.FileWriter("summary_logs", sess.graph)

#Now we can define and run the actual training loop, like this:
# Initialize reporting variables

inital_cost = 0
diff = 1
epoch_vals = []
accuracy_vals = []
costs = []
```

```python
# Training epochs
for i in range(num_epochs):
    if i > 1 and diff < .0001:
        print("change in cost %g; convergence."%diff)
        break

    else:
        # Run training step
        step = sess.run(model_train, feed_dict={input_values:
train_input_values, output_values: train_target_values})

        # Report some stats evert 10 epochs
        if i % 10 == 0:
            # Add epoch to epoch_values
            epoch_vals.append(i)

            # Generate the accuracy stats of the model
            train_accuracy, new_cost = sess.run([model_accuracy,
model_cost], feed_dict={input_values: train_input_values, output_values:
train_target_values})

            # Add accuracy to live graphing variable
            accuracy_vals.append(train_accuracy)

            # Add cost to live graphing variable
            costs.append(new_cost)

            # Re-assign values for variables
            diff = abs(new_cost - inital_cost)
            cost = new_cost

            print("Training step %d, accuracy %g, cost %g, cost change
%g"%(i, train_accuracy, new_cost, diff))
```

输出如下。

```
Training step 0, accuracy 0.343434, cost 34.6022, cost change 34.6022
Training step 10, accuracy 0.434343, cost 30.3272, cost change 30.3272
Training step 20, accuracy 0.646465, cost 28.3478, cost change 28.3478
Training step 30, accuracy 0.646465, cost 26.6752, cost change 26.6752
Training step 40, accuracy 0.646465, cost 25.2844, cost change 25.2844
Training step 50, accuracy 0.646465, cost 24.1349, cost change 24.1349
Training step 60, accuracy 0.646465, cost 23.1835, cost change 23.1835
```

```
Training step 70, accuracy 0.646465, cost 22.3911, cost change 22.3911
Training step 80, accuracy 0.646465, cost 21.7254, cost change 21.7254
Training step 90, accuracy 0.646465, cost 21.1607, cost change 21.1607
Training step 100, accuracy 0.666667, cost 20.677, cost change 20.677
Training step 110, accuracy 0.666667, cost 20.2583, cost change 20.2583
Training step 120, accuracy 0.666667, cost 19.8927, cost change 19.8927
Training step 130, accuracy 0.666667, cost 19.5705, cost change 19.5705
Training step 140, accuracy 0.666667, cost 19.2842, cost change 19.2842
Training step 150, accuracy 0.666667, cost 19.0278, cost change 19.0278
Training step 160, accuracy 0.676768, cost 18.7966, cost change 18.7966
Training step 170, accuracy 0.69697, cost 18.5867, cost change 18.5867
Training step 180, accuracy 0.69697, cost 18.3951, cost change 18.3951
Training step 190, accuracy 0.717172, cost 18.2191, cost change 18.2191
Training step 200, accuracy 0.717172, cost 18.0567, cost change 18.0567
Training step 210, accuracy 0.737374, cost 17.906, cost change 17.906
Training step 220, accuracy 0.747475, cost 17.7657, cost change 17.7657
Training step 230, accuracy 0.747475, cost 17.6345, cost change 17.6345
Training step 240, accuracy 0.757576, cost 17.5113, cost change 17.5113
Training step 250, accuracy 0.787879, cost 17.3954, cost change 17.3954
Training step 260, accuracy 0.787879, cost 17.2858, cost change 17.2858
Training step 270, accuracy 0.787879, cost 17.182, cost change 17.182
Training step 280, accuracy 0.787879, cost 17.0834, cost change 17.0834
Training step 290, accuracy 0.787879, cost 16.9895, cost change 16.9895
Training step 300, accuracy 0.79798, cost 16.8999, cost change 16.8999
Training step 310, accuracy 0.79798, cost 16.8141, cost change 16.8141
Training step 320, accuracy 0.79798, cost 16.732, cost change 16.732
Training step 330, accuracy 0.79798, cost 16.6531, cost change 16.6531
Training step 340, accuracy 0.808081, cost 16.5772, cost change 16.5772
Training step 350, accuracy 0.818182, cost 16.5041, cost change 16.5041
Training step 360, accuracy 0.838384, cost 16.4336, cost change 16.4336
Training step 370, accuracy 0.838384, cost 16.3655, cost change 16.3655
Training step 380, accuracy 0.838384, cost 16.2997, cost change 16.2997
Training step 390, accuracy 0.838384, cost 16.2359, cost change 16.2359
Training step 400, accuracy 0.848485, cost 16.1741, cost change 16.1741
Training step 410, accuracy 0.848485, cost 16.1141, cost change 16.1141
Training step 420, accuracy 0.848485, cost 16.0558, cost change 16.0558
Training step 430, accuracy 0.858586, cost 15.9991, cost change 15.9991
Training step 440, accuracy 0.858586, cost 15.944, cost change 15.944
Training step 450, accuracy 0.858586, cost 15.8903, cost change 15.8903
Training step 460, accuracy 0.868687, cost 15.8379, cost change 15.8379
Training step 470, accuracy 0.878788, cost 15.7869, cost change 15.7869
Training step 480, accuracy 0.878788, cost 15.7371, cost change 15.7371
Training step 490, accuracy 0.878788, cost 15.6884, cost change 15.6884
```

```
Training step 500, accuracy 0.878788, cost 15.6409, cost change 15.6409
Training step 510, accuracy 0.878788, cost 15.5944, cost change 15.5944
Training step 520, accuracy 0.878788, cost 15.549, cost change 15.549
Training step 530, accuracy 0.888889, cost 15.5045, cost change 15.5045
Training step 540, accuracy 0.888889, cost 15.4609, cost change 15.4609
Training step 550, accuracy 0.89899, cost 15.4182, cost change 15.4182
Training step 560, accuracy 0.89899, cost 15.3764, cost change 15.3764
Training step 570, accuracy 0.89899, cost 15.3354, cost change 15.3354
Training step 580, accuracy 0.89899, cost 15.2952, cost change 15.2952
Training step 590, accuracy 0.909091, cost 15.2558, cost change 15.2558
Training step 600, accuracy 0.909091, cost 15.217, cost change 15.217
Training step 610, accuracy 0.909091, cost 15.179, cost change 15.179
Training step 620, accuracy 0.909091, cost 15.1417, cost change 15.1417
Training step 630, accuracy 0.909091, cost 15.105, cost change 15.105
Training step 640, accuracy 0.909091, cost 15.0689, cost change 15.0689
Training step 650, accuracy 0.909091, cost 15.0335, cost change 15.0335
Training step 660, accuracy 0.909091, cost 14.9987, cost change 14.9987
Training step 670, accuracy 0.909091, cost 14.9644, cost change 14.9644
Training step 680, accuracy 0.909091, cost 14.9307, cost change 14.9307
Training step 690, accuracy 0.909091, cost 14.8975, cost change 14.8975
```

现在，看看训练的模型如何在 iris 数据集上执行，因此根据测试集测试训练的模型。

```
# test the model against the test set
print("final accuracy on test set: %s" %str(sess.run(model_accuracy,
feed_dict={input_values: test_input_values,
output_values: test_target_values})))
```

输出如下。

```
final accuracy on test set: 0.9
```

在测试集上获得了 0.9 的准确度，这非常好，我们可以通过更改迭代次数来尝试获得更好的结果。

5.8 总结

本章讨论了神经网络并解释了为什么需要多层神经网络，还介绍了 TensorFlow 计算图模型，其中包含一些基本示例，如线性回归和逻辑回归。

接下来，本书将介绍更高级的示例，并演示 TensorFlow 如何用于实现手写字符识别等，同时还将介绍在传统机器学习中取代特征工程的架构工程的核心理念。

第 6 章 深度前馈神经网络——实现数字分类

前馈神经网络（Feed-forward Neural Network，FNN）是一种特殊类型的神经网络，其中神经元之间的链接/连接不构成环。因此，它与本书后续将要研究的神经网络中的其他结构（如循环类型神经网络）不同。FNN 是一种广泛使用的架构，而且它是第一个神经网络，并且是类型最简单的。

本章将会介绍一种典型的 FNN 架构，并使用 TensorFlow 的库来实现它。在讲完了这些概念之后，本章会给出一个数字分类的实际例子。这个例子的问题是：给定一组包含手写数字的图像，如何将这些图像分为 10 个不同的类别（0~9）？

本章将包含以下几个主题。

- 隐藏单元与架构设计。
- MNIST 数据集分析。
- 数字分类——构建与训练模型。

6.1 隐藏单元与架构设计

在下一节中，我们将会回顾人工神经网络：它可以很好地完成分类任务，比如对手写数字进行分类。

假设现在的网络如图 6.1 所示。

如前所述，该网络中最左边的一层称为**输入层**，该层内的神经元称为**输入神经元**，最右边一层称为输出层，它包含了输出神经元。在本例中，输出层只包含单个输出神经元。中间层称为**隐藏层**，因为该层中的神经元既不是输入，也不是输出。隐藏这个词可能听起来有点神秘——第一次听到这个词的时候，我觉得它一定具有某种深刻的哲理或者

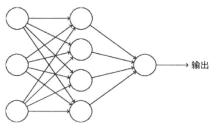

图 6.1 含有一层隐藏层的简单前馈神经网络

数学意义，但事实上它真的表示既不是输入也不是输出，没有什么别的意思。前面这个网络仅仅只有一个隐藏层，但是有些网络有多个隐藏层。例如，图 6.2 所示这个 4 层的网络包含两个隐藏层。

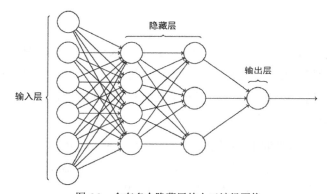

图 6.2 含有多个隐藏层的人工神经网络

这种描述输入层、隐藏层和输出层的架构显得非常简单。通过一个实际的例子来查看特定的手写图像是否包含数字 9。

首先，需要将输入图像的像素提供给输入层，例如，在 MNIST 数据集中，图像是单色图像。因为每一幅图都是 28×28 像素大小的，所以为了接收输入图像在输入层中需要 28×28=784 个神经元。

在输出层中，只需要一个神经元，它产生该图像是否包含数字 9 的概率（或分数）。例如，输出值大于 0.5 表示该图像包含数字 9；输出值小于 0.5 代表输入图像中不包含数字 9。

这种将一层的输出作为下一层的输入的网络称为前馈神经网络（FNN）。层间的这种顺序性意味着其中并没有循环。

6.2 MNIST 数据集分析

本节将动手为手写图像实现一个分类器。这种实现可以被视为神经网络中的"Hello world!"。

MNIST 是一个广泛用于对机器学习技术进行基准测试的数据集。该数据集包含一组手写数字，如图 6.3 所示。

图 6.3　MNIST 数据集中的样本数字

除了包含手写图像之外，该数据集还包含其对应的标签。

本节将会在这些图像上训练一个基本的模型，目标是确定输入图像中哪个数字是手写的。

另外，我们会发现这里使用几行代码就能够完成这个分类任务，但是此实现背后的想法是理解构建神经网络解决方案的基本要点。同时，在该实现中本节将介绍神经网络的主要概念。

MNIST 数据

MNIST 数据托管在 YannLecun 的网站上。幸运的是，TensorFlow 提供了一些辅助函数来下载该数据集，因此本节从使用以下两行代码下载该数据集开始。

```
from tensorflow.examples.tutorials.mnist import input_data
mnist_dataset = input_data.read_data_sets("MNIST_data/", one_hot=True)
```

MNIST 数据分为 3 部分：55000 个训练数据样本（mnist.train）、10000 个测试数据样本（mnist.test）和 5000 个验证数据样本（mnist.validation）。这种划分方式非常重要，在机器学习中重要的是：拥有在训练过程中没有使用过的不同数据，以确保最终模型所学到的东西确实是可以泛化的。

如前所述，每一个 MNIST 样本都有两部分：手写数字图像及其对应的标签。训练集和测试集均包含：图像及其对应的标签。例如，训练图像是 mnist.train.images，训练标签是 mnist.train.labels。

每一幅图像都是 28×28 像素的，并且可以用一个矩阵来表示，如图 6.4 所示。

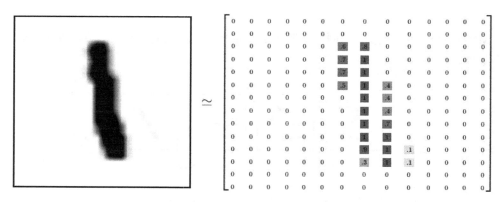

图 6.4　MNIST 数字的矩阵表示（像素强度值）

为了将这个像素值矩阵输入神经网络的输入层，需要将该矩阵展平为一个 784 维的向量。因此，数据集的最终形状将是一个 784 维的向量空间。

于是 mnist.train.images 是一个形状为（55000，784）的张量，如图 6.5 所示。第一维是图像序列的索引，第二维是每幅图像中每个像素的索引。该张量中的每一项都是某幅特定图像中某一特定像素介于 0 到 1 的像素强度值。

如前所述，该数据集中的每一幅图像都有其对应的标签，范围是 0～9。

出于该实现的目的，这里将标签编码为独热向量（one-hot vector）。独热向量是指除了这个向量所代表的数字的索引位是 1 外，其余位全为 0 的向量。例如，3 编码为独热向量就是[0,0,0,1,0,0,0,0,0,0]。这样一来，mnist.train.labels 就是一个（55000，10）大小的浮点数组，如图 6.6 所示。

图 6.5　MNIST 数据分析　　　　　图 6.6　MNIST 数据分析

6.3 数字分类——构建与训练模型

现在，可以继续构建模型了。现在数据集中有 0~9 这 10 个类别，目标是将任何一幅输入图像划分到其中一类。通常情况下会给出输入图像的一个硬性决定，指出它到底属于其中的哪一类。这里不这样做，而会给出一个含有 10 个可能值的向量（因为有 10 个可能的类别）。它们表示输入图像属于 0~9 中每一类的可能性。

例如，假设为模型提供特定图像。该模型有 70% 的概率确定此图像为 9，有 10% 的概率确定此图像为 8，依次类推。所以，这里将会使用 softmax 回归，它将产生 0~1 的值。

softmax 回归有两个步骤：首先将输入属于确定类的证据加在一起，然后将该特征转换为概率。

为了计算给定图像属于特定类的证据，这里对像素强度求加权和。如果具有高强度值的像素证明图像不属于某一类，那么权重就是负值；如果该像素证明图像属于某一类，那么权重就是正值。

图 6.7 所示模型为每个类别学习到的权重。

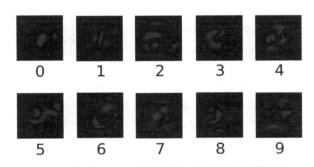

图 6.7 模型为 MNIST 数据中每一个类别学习到的权重

这里还增加了一些名为**偏差**的额外证据项。基本上，我们希望能够指出某些事物更可能与输入无关。于是，给定输入 x，其属于类别 i 的证据项就是：

$$\text{evidence}_i = \sum_j W_{i,j} x_j + b_i$$

其中，W_i 是权重；b_i 是类别 i 的偏差；j 是在输入图像 x 的像素上求和时的索引。

然后，使用 softmax 函数将证据值转换为预测概率 y。

$$y = \text{softmax}(\text{evidence})$$

这里，softmax 是一个激活或链接函数，它将线性函数的输出变换为期望的形式。在本例中，它就是 10 个类别（因为这里有 0～9 这 10 个可能的类别）上的一个概率分布。我们可以将它视为把证据值转换为输入属于每一类别的概率。它定义为：

$$\text{softmax}(\text{evidence}) = \text{normalize}(\exp(\text{evidence}))$$

如果展开这个等式，就会得到：

$$\text{softmax}(\text{evidence})_i = \frac{\exp(\text{evidence}_i)}{\sum_j \exp(\text{evidence}_j)}$$

但是，用第一种方式考虑 softmax 通常更有帮助：对输入求幂，然后对它们进行归一化。求幂意味着增加一个单位的证据可以指数级增加任何假设的权重。相反，减少一个单位的证据意味着某个假设的权重减小。没有权重是零值或者负值的假设。softmax 然后对这些权重进行归一化，于是权重的和为 1，形成了一个有效的概率分布。

softmax 回归的图解如图 6.8 所示。当然，图中有更多的 x。对于每个输出，计算 x 的加权和，加上偏差项，然后应用 softmax 回归。

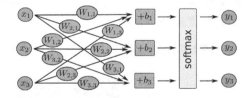

图 6.8　softmax 回归的可视化

如果将它写成等式，如图 6.9 所示。

可以使用矢量符号来表示此过程。这意味着将它变成矩阵乘法和向量加法。这对于计算效率和可读性都是非常有帮助的，如图 6.10 所示。

更简洁的写法如下。

$$y = \text{softmax}(\boldsymbol{W}x + b)$$

图 6.9　softmax 回归的等式表示　　　图 6.10　softmax 回归等式的向量化表示

现在，考虑将它变为 TensorFlow 能够使用的东西。

6.3.1 分析数据

现在我们继续实现分类器。首先，导入此实现需要用到的包。

```
import tensorflow as tf
import matplotlib.pyplot as plt
import numpy as np
import random as ran
```

然后，定义一些辅助函数，使得能够在下载的原始数据集中进行子集化。

```
#Define some helper functions
# to assign the size of training and test data we will take from MNIST dataset
def train_size(size):
    print ('Total Training Images in Dataset = ' + str(mnist_dataset.train.images.shape))
    print ('############################################')
    input_values_train = mnist_dataset.train.images[:size,:]
    print ('input_values_train Samples Loaded = ' + str(input_values_train.shape))
    target_values_train = mnist_dataset.train.labels[:size,:]
    print ('target_values_train Samples Loaded = ' + str(target_values_train.shape))
    return input_values_train, target_values_train

def test_size(size):
    print ('Total Test Samples in MNIST Dataset = ' + str(mnist_dataset.test.images.shape))
    print ('############################################')
    input_values_test = mnist_dataset.test.images[:size,:]
    print ('input_values_test Samples Loaded = ' + str(input_values_test.shape))
    target_values_test = mnist_dataset.test.labels[:size,:]
    print ('target_values_test Samples Loaded = ' + str(target_values_test.shape))
    return input_values_test, target_values_test
```

此外，我们还将定义两个辅助函数，用来显示数据集中的特定数字，或者显示图像子集展开的版本。

```
#Define a couple of helper functions for digit images visualization
def visualize_digit(ind):
    print(target_values_train[ind])
```

```
    target = target_values_train[ind].argmax(axis=0)
    true_image = input_values_train[ind].reshape([28,28])
    plt.title('Sample: %d Label: %d' % (ind, target))
    plt.imshow(true_image, cmap=plt.get_cmap('gray_r'))
    plt.show()

def visualize_mult_imgs_flat(start, stop):
    imgs = input_values_train[start].reshape([1,784])
    for i in range(start+1,stop):
        imgs = np.concatenate((imgs,
input_values_train[i].reshape([1,784])))
    plt.imshow(imgs, cmap=plt.get_cmap('gray_r'))
    plt.show()
```

现在,可以开始讨论业务并且使用数据集了。所以,我们将会定义希望从原始数据集中加载的训练和测试样本。

下面开始构建和训练模型。首先,根据需要加载的训练和测试样本来定义变量。目前,该程序会加载所有数据,但是后续可能会更改此值以节省资源。

```
input_values_train, target_values_train = train_size(55000)
```

输出如下。

```
Total Training Images in Dataset = (55000, 784)
##############################################
input_values_train Samples Loaded = (55000, 784)
target_values_train Samples Loaded = (55000, 10)
```

现在有一个包含55000个手写数字样本的训练集。其中,每个样本都是28×28像素的图像,被展平为784维的向量,而且它还有对应的独热编码格式的标签。

这里 target_values_train 数据是所有 input_values_train 样本的关联标签。在下面这个例子中,图6.11 所示的数组表示数字 7 的独热编码格式。

标签	0	1	2	3	4	5	6	7	8	9
数组	[0,	0,	0,	0,	0,	0,	0,	1,	0,	0]

图6.11 数字 7 的独热编码

那么,通过以下代码从数据集中随机抽出一张图进行可视化,来看一下它是什么样的。

```
visualize_digit(ran.randint(0, input_values_train.shape[0]))
```

这里会用到前面定义的辅助函数来从数据集中显示一个随机的数字，输出如图 6.12 所示。

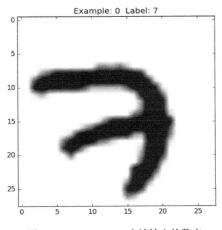

图 6.12　display_digit 方法输出的数字

还可以在以下代码中使用之前定义的辅助函数来可视化一堆展平的图像。

```
visualize_mult_imgs_flat(0,400)
```

因为展平的向量中的每一个值代表像素强度，所以可视化这些像素后的图形如图 6.13 所示。

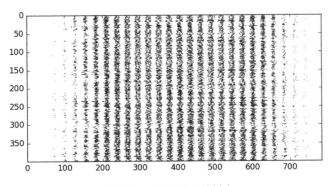

图 6.13　前 400 个训练样本

6.3.2 构建模型

到目前为止,我们尚未开始给此分类器构建计算图。现在从创建会话变量开始,该变量将负责执行将要构建的计算图。

```
sess = tf.Session()
```

接下来,将定义模型的占位符,这些占位符将用于给计算图提供数据。

```
input_values = tf.placeholder(tf.float32, shape=[None, 784])
```

当在占位符的第一维中指定 None 时,这意味着可以根据需要为占位符提供尽可能多的样本。在这种情况下,占位符可以接收任意数量的样本,其中每个样本具有 784 个值。

现在需要定义另一个占位符来提供图像标签。另外,之后会用此占位符来比较模型预测和图像的真实标签。

```
output_values = tf.placeholder(tf.float32, shape=[None, 10])
```

接下来,将会定义 weights 和 bias。这两个变量将是网络中的可训练参数,而且它们是在未知数据上做预测所需要的仅有的两个变量。

```
weights = tf.Variable(tf.zeros([784,10]))
biases = tf.Variable(tf.zeros([10]))
```

可以将这些 weights 视为每个数字的 10 个备忘单。这有点类似于教师是如何使用备忘单对多项选择考试进行评分的。

现在可以定义 softmax 回归了,它是本节的分类器函数。这种特定的分类器称为**多项逻辑回归**,我们通过将数字的展平版本乘以权重然后加上偏差进行预测。

```
softmax_layer = tf.nn.softmax(tf.matmul(input_values,weights) + biases)
```

首先,忽略 softmax 本身,仅仅查看 softmax 函数里面的内容。matmul 是 TensorFlow 中用于矩阵乘法的函数。如果读者了解矩阵乘法,就会明白这样计算是合适的,并且 $xW+b$ 将会得到一个训练样本数(m)×类别数(n)大小的矩阵,如图 6.14 所示。

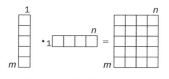

图 6.14 简单的矩阵乘法

可以通过计算 softmax_layer 来确认它。

```
print(softmax_layer)
Output:
Tensor("Softmax:0", shape=(?, 10), dtype=float32)
```

现在，用之前定义的计算图开始实验，使用训练集中的 3 个样本，看一下它是如何工作的。要执行计算图，需要用到之前定义的会话变量，并且需要使用 tf.global_variables_initializer() 来初始化变量。

现在继续，只给计算图提供 3 个样本。

```
input_values_train, target_values_train = train_size(3)
sess.run(tf.global_variables_initializer())
#If using TensorFlow prior to 0.12 use:
#sess.run(tf.initialize_all_variables())
print(sess.run(softmax_layer, feed_dict={input_values:
input_values_train}))
```

输出如下。

```
[[ 0.1 0.1 0.1 0.1 0.1 0.1 0.1 0.1 0.1 0.1]
 [ 0.1 0.1 0.1 0.1 0.1 0.1 0.1 0.1 0.1 0.1]
 [ 0.1 0.1 0.1 0.1 0.1 0.1 0.1 0.1 0.1 0.1]]
```

在这里，可以看到模型对提供给它的 3 个训练样本的预测。目前，模型对于这个任务没有学到任何东西，因为还没有经历过训练过程，所以它的输出表明对于输入样本每一个数字正确分类的概率均为 10%。

如前所述，softmax 是一个将输出压缩到 0～1 的激活函数，并且 softmax 的 TensorFlow 实现确保了单个输入样本属于所有类别的概率之和为 1。

下面使用 TensorFlow 的 softmax 函数进行一些实验。

```
sess.run(tf.nn.softmax(tf.zeros([4])))
sess.run(tf.nn.softmax(tf.constant([0.1, 0.005, 2])))
```

输出如下。

```
array([0.11634309, 0.10579926, 0.7778576 ], dtype=float32)
```

接下来,需要给模型定义损失函数,它将衡量分类器在尝试为输入图像划分类别时的好坏。模型的准确性是通过比较数据集中的真实值与从模型中得到的预测值得到的。

目标是减少真实值与预测值之间的任何分类错误。

交叉熵定义为:

$$H_{y'}(y) = -\sum_{i} y'_i \log(y_i)$$

其中,y 是预测的概率分布;y' 是真实分布(带有数字标签的独热向量)。

在广义上,交叉熵衡量了模型预测对于描述真实输入的无效性。

可以这样实现交叉熵函数。

```
model_cross_entropy = tf.reduce_mean(-tf.reduce_sum(output_values *
tf.log(softmax_layer), reduction_indices=[1]))
```

此函数对于从 softmax_layer 中获得的所有预测(值范围是 0~1)取对数,并将它们与样本的真实值 output_values 逐一相乘(参见维基百科)。如果 log 函数作用在每一个接近于 0 的值上,那么得到的结果将是一个很大的负数(-np.log(0.01) = 4.6);如果 log 函数作用在接近于 1 的数上,那么结果将是一个很小的负数(-np.log(0.99) = 0.1)。log 函数的图像如图 6.15 所示。

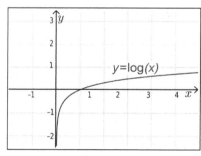

图 6.15 函数 $y=\log(x)$ 的可视化

如果预测确实不准确,那么算法基本上会使用非常大的数字来惩罚分类器;如果预测确实准确,那么惩罚也会非常小。

下面是一个 softmax 预测的简单 Python 例子,它非常确信数字是 3。

```
j = [0.03, 0.03, 0.01, 0.9, 0.01, 0.01, 0.0025, 0.0025, 0.0025, 0.0025]
```

再创建一个标签为 3 的数组作为标准,来和 softmax 函数进行对比。

```
K = [0,0,0,1,0,0,0,0,0,0]
```

你能够猜到损失函数给出了什么值吗?你能够理解对 j 取对数后是如何用一个很大的负值来惩罚错误答案吗?试试用以下代码来理解。

```
-np.log(j)
-np.multiply(np.log(j),k)
```

这将返回 9 个 0 值,以及 1 个 0.1053。当将它们全部加起来的时候,认为这是一个很好的预测。注意当对实际的 2 进行相同的预测时会发生什么。

```
k = [0,0,1,0,0,0,0,0,0,0]
np.sum(-np.multiply(np.log(j),k))
```

现在,cross_entropy 函数给出了结果 4.6051,这表示严重的惩罚,这是一个预测不佳的结果。它被严重惩罚的原因是分类器十分确信样本是 3,但实际上样本是数字 2。

接下来,开始训练分类器。为了训练它,必须要给 W 和 b 找到合适的值,以尽可能降低损失。

现在可以根据需要为训练分配自定义变量。在整个过程中任何像下面这样的值都可以随意更改。事实上,算法也鼓励这样做。首先,使用这些值可以监测到使用很少的训练样本或者使用很高/低的学习率时会发生什么。

```
input_values_train, target_values_train = train_size(5500)
input_values_test, target_values_test = test_size(10000)
learning_rate = 0.1
num_iterations = 2500
```

现在可以初始化所有变量,以便可以在 TensorFlow 图中使用它们。

```
init = tf.global_variables_initializer()
#If using TensorFlow prior to 0.12 use:
#init = tf.initialize_all_variables()
sess.run(init)
```

接下来,需要使用梯度下降算法来训练分类器。所以首先定义训练方法和一些用来衡量模型准确性的变量。变量 train 将会使用选定的学习率来实现梯度下降优化器,以最小化模型损失函数 model_cross_entropy。

```
train
=tf.train.GradientDescentOptimizer(learning_rate).minimize(model_cross_entropy)
model_correct_prediction = tf.equal(tf.argmax(softmax_layer,1),
tf.argmax(output_values,1))
model_accuracy = tf.reduce_mean(tf.cast(model_correct_prediction,
tf.float32))
```

6.3.3 训练模型

现在,定义一个迭代 num_iterations 次的循环。对于每次循环,程序都会执行训练,使用 feed_dict 从 input_values_train 和 target_values_train 获取输入值。

出于计算准确性,我们将会使用 input_values_test 中的未知数据来测试模型。

```
for i in range(num_iterations+1):
    sess.run(train, feed_dict={input_values: input_values_train,
output_values: target_values_train})
    if i%100 == 0:
        print('Training Step:' + str(i) + ' Accuracy = ' +
str(sess.run(model_accuracy, feed_dict={input_values: input_values_test,
output_values: target_values_test})) + ' Loss = ' +
str(sess.run(model_cross_entropy, {input_values: input_values_train,
output_values: target_values_train})))
```

输出如下。

```
Training Step:0 Accuracy = 0.5988 Loss = 2.1881988
Training Step:100 Accuracy = 0.8647 Loss = 0.58029664
Training Step:200 Accuracy = 0.879 Loss = 0.45982164
Training Step:300 Accuracy = 0.8866 Loss = 0.40857208
Training Step:400 Accuracy = 0.8904 Loss = 0.37808096
Training Step:500 Accuracy = 0.8943 Loss = 0.35697535
Training Step:600 Accuracy = 0.8974 Loss = 0.34104997
Training Step:700 Accuracy = 0.8984 Loss = 0.32834956
Training Step:800 Accuracy = 0.9 Loss = 0.31782663
Training Step:900 Accuracy = 0.9005 Loss = 0.30886236
Training Step:1000 Accuracy = 0.9009 Loss = 0.3010645
Training Step:1100 Accuracy = 0.9023 Loss = 0.29417014
Training Step:1200 Accuracy = 0.9029 Loss = 0.28799513
Training Step:1300 Accuracy = 0.9033 Loss = 0.28240603
Training Step:1400 Accuracy = 0.9039 Loss = 0.27730304
Training Step:1500 Accuracy = 0.9048 Loss = 0.27260992
Training Step:1600 Accuracy = 0.9057 Loss = 0.26826677
Training Step:1700 Accuracy = 0.9062 Loss = 0.2642261
Training Step:1800 Accuracy = 0.9061 Loss = 0.26044932
Training Step:1900 Accuracy = 0.9063 Loss = 0.25690478
Training Step:2000 Accuracy = 0.9066 Loss = 0.2535662
```

```
Training Step:2100 Accuracy = 0.9072 Loss = 0.25041154
Training Step:2200 Accuracy = 0.9073 Loss = 0.24742197
Training Step:2300 Accuracy = 0.9071 Loss = 0.24458146
Training Step:2400 Accuracy = 0.9066 Loss = 0.24187621
Training Step:2500 Accuracy = 0.9067 Loss = 0.23929419
```

请注意在将要结束时损失值仍然在下降，但是准确率略有下降。这表明仍然可以最大限度地减少损失，从而最大限度地提高模型在训练数据上的准确性，但是这可能无法帮助我们在用于衡量准确性的测试数据上做预测，这也称为**过拟合**（泛化性不强）。若使用默认设置，准确性大约为91%。如果要作弊以获得94%的准确性，可以将测试样本设置为100个。这表明没有足够的测试样本可能会导致有偏差的准确性。

注意，这是计算分类器性能的一种很不正确的方式。当然，这里出于学习和实验目的故意这样做。理想情况下，当使用大型数据集训练时，每一次使用其中小批量训练数据进行训练，而不是一次使用所有的数据。

接下来就是有趣的部分了。既然已经计算出了权重备忘单，接下来就可以使用以下代码来构建一个图。

```
for i in range(10):
    plt.subplot(2, 5, i+1)
    weight = sess.run(weights)[:,i]
    plt.title(i)
    plt.imshow(weight.reshape([28,28]), cmap=plt.get_cmap('seismic'))
    frame = plt.gca()
    frame.axes.get_xaxis().set_visible(False)
    frame.axes.get_yaxis().set_visible(False)
```

以上代码构建出的图如图6.16所示。

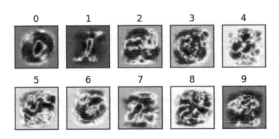

图6.16 权重0~9的可视化

图6.16显示了模型学习到的0~9的权重，这也是分类器最重要的一面。机器学习

涉及的大量工作就是要弄清楚最优权重是什么。一旦根据某种优化标准计算出了最优权重，就能得到备忘录，并且可以使用学习到的权重轻松找到问题的答案。

学习的模型通过比较输入的数字样本和图中两种颜色的权重的相似度或差异性来做出预测。不同颜色代表的匹配度不一样。在计算机屏幕上，红色越深，表示匹配越好；白色表示中性；蓝色表示不匹配。

现在，可以使用备忘单来看看模型的表现如何。

```
input_values_train, target_values_train = train_size(1)
visualize_digit(0)
```

输出如下。

```
Total Training Images in Dataset = (55000, 784)
################################################
input_values_train Samples Loaded = (1, 784)
target_values_train Samples Loaded = (1, 10)
[0. 0. 0. 0. 0. 0. 0. 1. 0. 0.]
```

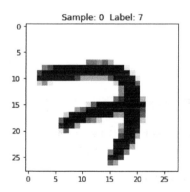

下面看 softmax 预测器。

```
answer = sess.run(softmax_layer, feed_dict={input_values:
input_values_train})
print(answer)
```

前面的代码将会给出一个 10 维的向量，每一列包含一个概率值。

```
[[2.1248012e-05 1.1646927e-05 8.9631692e-02 1.9201526e-02 8.2086492e-04
  1.2516821e-05 3.8538201e-05 8.5374612e-01 6.9188857e-03 2.9596921e-02]]
```

可以使用 argmax 函数来找到输入图像最有可能属于的类别。

```
answer.argmax()
```

输出如下。

7

现在，我们从网络中得到了一个正确的分类。

使用我们的知识来定义一个辅助函数，使它能够从数据集中随机选择一幅图像，并用该图像来测试模型。

```
def display_result(ind):
    # Loading a training sample
    input_values_train = mnist_dataset.train.images[ind,:].reshape(1,784)
    target_values_train = mnist_dataset.train.labels[ind,:]
    # getting the label as an integer instead of one-hot encoded vector
    label = target_values_train.argmax()
    # Getting the prediction as an integer
    prediction = sess.run(softmax_layer, feed_dict={input_values: input_values_train}).argmax()
    plt.title('Prediction: %d Label: %d' % (prediction, label))
    plt.imshow(input_values_train.reshape([28,28]), cmap=plt.get_cmap('gray_r'))
    plt.show()
```

现在来尝试一下该辅助函数。

```
display_result(ran.randint(0, 55000))
```

输出如下。

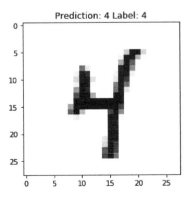

再次得到了一个正确的分类。

6.4 总结

本章主要讲述了如何使用 FNN 完成数字分类任务,并回顾了神经网络环境中使用的一些术语。

接下来,我们将会使用一些现代最佳实践和技巧来建立一个复杂版本的数字分类模型,以增强模型的性能。

第 7 章 卷积神经网络

在数据科学中，**卷积神经网络**（Convolutional NeuralNetwork，CNN）是一类特定的深度学习架构，它利用卷积运算来挖掘输入图像中的相关的可解释特征。CNN 层之间像前馈神经网络一样互相连接，同时它利用卷积操作来模拟人类识别物体时大脑的工作原理。单个皮层神经元只对空间中某个受限区域的刺激做出反应，这称为感受野。特别地，生物医学成像曾一度是一个很有挑战性的问题，但在本章中，读者将看到如何利用 CNN 来发掘图像中的模式。

本章将包含以下主题。

- 卷积运算。
- 动机。
- CNN 的不同层。
- CNN 基础示例——MNIST 手写数字分类。

7.1 卷积运算

CNN 在计算机视觉领域里广泛应用，并且它们表现得比前人所使用的大多数传统计算机视觉技术要出色。CNN 将著名的卷积运算和神经网络结合在一起，因此叫作卷积神经网络。因此，在深入讨论 CNN 的神经网络部分之前，本书将先介绍卷积运算并看看它是怎么工作的。

卷积运算的主要目的是从图像中提取信息或特征。任何图像都可以看作一个数值矩阵，而矩阵中一组特定的数值可以构成一个特征。卷积运算的目的是扫描这个矩阵，并尝试为图像挖掘相关的或可解释特征。例如，考虑一个 5×5 的图像，它的对应灰度（或者说像素值），用 0 和 1 来表示，如图 7.1 所示。

再考虑图 7.2 所示的 3×3 的矩阵。

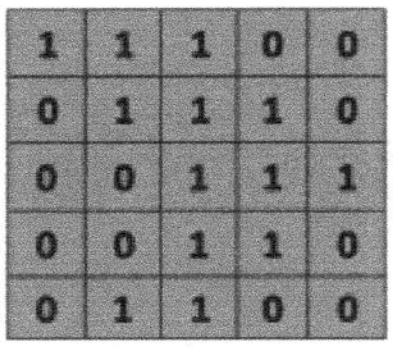

图 7.1　5×5 的像素值矩阵

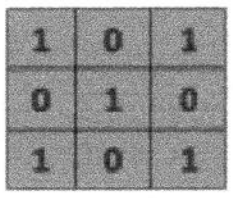

图 7.2　3×3 的像素值矩阵

如图 7.3（a）～（i）所示，可以用 3×3 的矩阵对 5×5 的图像进行卷积。

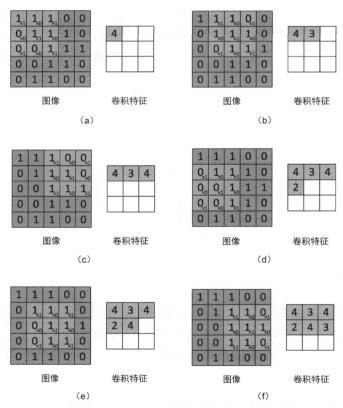

图 7.3　卷积运算。输出矩阵称为卷积特征或卷积映射

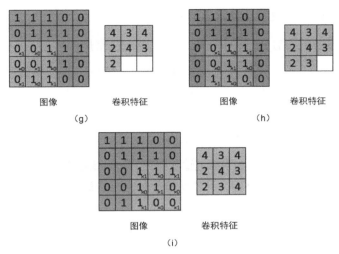

图 7.3 卷积运算。输出矩阵称为卷积特征或卷积映射（续）

图 7.3 可以概括为以下内容，为了利用 3×3 卷积核对原有的 5×5 图像进行卷积，需要进行以下操作。

- 用 3×3 矩阵对原有的 5×5 图像进行扫描，并且每次移动一个像素（步长）。
- 对于 3×3 图像中的不同位置，用 3×3 矩阵中的每一个元素乘以 5×5 矩阵中的对应像素值。
- 把这些对应元素相乘的结果加在一起得到一个数值，这个数值构成了右侧输出矩阵里的单个数值。

> 正如我们在图 7.3 中看到的，3×3 矩阵在每一步时仅仅在 5×5 原图像的一部分上进行操作，或者说每次只能看到原图像的一部分。

在前面的叙述里，CNN 中的术语描述如下。

- 3×3 矩阵称为**核**（kernel）、**特征探测器**（feature detector）或**过滤器**（filter）。
- 包含对应元素相乘结果的右侧输出矩阵称为**特征映射**（feature map）。

因为特征映射是基于核和原输入图像上对应像素之间的对应元素相乘获得的，所以每次改变核或者过滤器的数值都会产生不同的特征映射。

于是，读者可能认为他自己需要在卷积神经网络的训练中弄清过滤器的数值，但实际情况并不是那样的。CNN 在学习过程中自动弄清了这些数字。因此，如果有更多的过滤器，就意味着可以从图像中挖掘更多的特征。

在开始学习下一节内容之前，先介绍 CNN 中一些常用的术语。

- **步长**（stride）：这个术语在前面简单提到过。通常，步长指的是特征探测器（或者过滤器）在输入矩阵的像素上移动的像素个数。例如，步长 1 表示在对输入图像做卷积时每次对过滤器移动 1 像素，步长 2 表示在对输入图像做卷积时每次对过滤器移动 2 像素。步长越大，生成的特征映射越小。
- **补零法**（Zero-padding）：如果想把输入图像的边沿像素也计算进来，那么过滤器有一部分就会落在输入图像外面。补零法通过在输入图像边沿上补零解决了这个问题。

7.2 动机

传统计算机视觉技术曾一度用来解决一些计算机视觉问题，例如目标检测和图像分割。这些传统计算机视觉技术表现得不错，但是它们离实际场景下的应用还有很大的距离，例如，应用在自动驾驶中。2012 年，AlexKrizhevsky 引入了 CNN，该方法在 ImageNet 比赛中将目标分类错误率从 26% 降低到了 25%，取得了巨大的突破。从那之后，CNN 就广泛地应用，并且提出了许多不同的变种。它在 ImageNet 比赛中已经击败了人类，呈现出了更低的错误分类率，如图 7.4 所示。

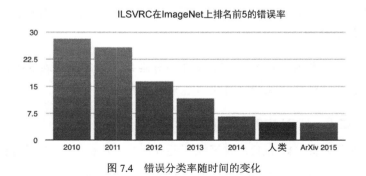

图 7.4　错误分类率随时间的变化

CNN 的应用

鉴于 CNN 在计算机视觉的不同领域乃至自然语言处理领域所取得的突破，很多公司已经将这种深度学习解决方案整合到了自己的计算机视觉回声系统中。例如，Google 在它们的图片搜索引擎中用到了该架构，而 Facebook 则用它来实现自动标注和其他工作，如图 7.5 所示。

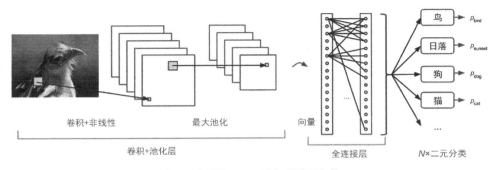

图 7.5　典型的 CNN 目标识别常用架构

CNN 取得这样的突破是源于它的架构，该架构直观地利用卷积操作来提取图像的特征。在后面的内容中，我们将看到这种方法和人脑的工作原理十分相像。

7.3　CNN 的不同层

如图 7.5 所示，一个典型的 CNN 架构包含了多个用于完成不同任务的层。本节将深入介绍它们的细节，并展示把它们按照特定的方法连接在一起后的好处，从而在计算机视觉领域取得如此大的突破。

7.3.1　输入层

这是 CNN 架构中的第一层，所有其后的卷积层和池化层都要求输入遵照特定的格式。输入变量是遵照下面形状的张量。

[batch_size, image_width, image_height, channels]

其中，batch_size 是从原始训练集中选取的随机样本（数量），在应用随机梯度下降法时使用；image_width 是网络的输入图像的宽度；image_height 是网络的输入图像的高度；channels 是输入图像颜色通道的数量，对于 RGB 图像通道数就是 3，对于二值图像就是 1。

举个例子，考虑著名的 MNIST 数据集。比如，要用 CNN 在这个数据集上做数字分类。

如果数据集像 MNIST 数据集一样由 28×28 像素的黑白图像组成，那么输入层所要求的形状如下所示。

[batch_size, 28, 28, 1]

要改变输入特征的形状，可以用下面的重塑操作。

```
input_layer = tf.reshape(feature["x"], [-1, 28, 28, 1])
```

 正如读者所见，这里将批尺寸（batch size）设置为-1，这表示这个数值的大小由特征里的输入值动态地决定。这样，我们就可以通过控制批尺寸对 CNN 模型进行微调了。

作为一个重塑操作的例子，假设把输入样本分为大小为 5 的批次，那么 feature["x"] 数组会包含输入图像的 3920 个数值，其中每个数值对应图像当中的一个像素。在本例中，输入层具有如下形状。

```
[5, 28, 28, 1]
```

7.3.2 卷积步骤

正如前文提到的，卷积步骤是因为卷积操作而取名的。加入卷积步骤的主要目的是从输入图像中提取特征，然后将它们输入线性分类器当中。

在自然图像中，特征可能出现在图像的任何位置。比如，边可能出现在图像的中间或者角落，因此堆叠大量卷积步骤的整个想法是能够在图像中的任何位置检测这些特征。

在 TensorFlow 里面很容易定义卷积步骤。比如，如果要对输入层使用带 ReLU 激活函数的 20 个 5×5 的过滤器，那么可以用下面这段代码来实现。

```
conv_layer1 = tf.layers.conv2d(
 inputs=input_layer,
 filters=20,
 kernel_size=[5, 5],
 padding="same",
 activation=tf.nn.relu)
```

conv2d 函数的第一个参数是在之前代码中定义的输入层，它具有合适的形状；第二个参数是过滤器参数，它指明了用在图像上的过滤器个数，过滤器数量越多，从输入图像中挖掘出来的特征也越多；第三个参数是核尺寸（kernel size），它表示过滤器或者特征检测器的尺寸。补齐（padding）参数（此处设置为"same"）指定了用补零法对输入图像边沿像素进行处理。最后一个参数指明了用在卷积操作的输出上的激活函数。

因此，在 MNIST 例子中，输入张量的形状是：

```
[batch_size, 28, 28, 1]
```

而卷积步骤的输出张量的形状是：

```
[batch_size, 28, 28, 20]
```

输出张量和输入张量有相同的维度，但是输出张量有 20 个通道，这表示对输入图像使用了 20 个过滤器。

7.3.3 引入非线性

在卷积步骤中，提到了要把卷积步骤的输出输入 ReLU 激活函数中来引入非线性，如图 7.6 所示。

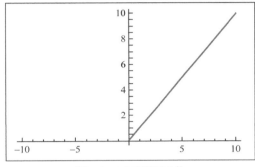

图 7.6　ReLU 激活函数

其中，输出=Max（0，输入）。

ReLU 激活函数把所有负的像素值替换为零。把卷积步骤的输出输入这个激活函数的主要目的是在输出图像中引入非线性，这对训练过程非常有用，因为数据通常都是非线性的。为了对 ReLU 激活函数的好处有个清楚的了解，可以看看图 7.7，它展示了卷积步骤的行输出以及它的修正版本。

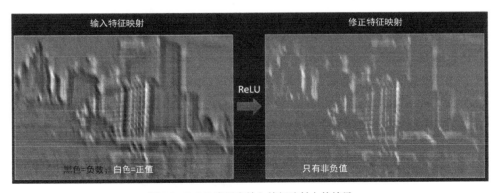

图 7.7　ReLU 应用在输入特征映射上的结果

7.3.4 池化步骤

学习过程中很重要的一步是池化步骤，它有时也称为下采样或者降采样。这一步骤旨在减少卷积步骤输出（特征映射）的维数。池化步骤的优点是能够减少特征映射的尺寸，同时在新缩减的版本里保留重要的信息。

图 7.8 通过一个 2×2 的过滤器以步长 2 在图像上扫描，同时在每个位置上使用最大化操作来展示这一步骤。这种池化操作称为**最大池化**（max pooling）。

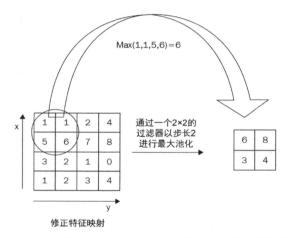

图 7.8　使用 2×2 窗口在修正特征映射（卷积和 ReLU 操作后获得）上
使用最大池化操作的一个例子（图片来源：textminingon line 网站）

可以用下面的代码将卷积步骤的输出连接到池化层。

```
pool_layer1 = tf.layers.max_pooling2d(inputs=conv_layer1, pool_size=[2, 2],strides=2)
```

池化层接收从卷积步骤得到的有以下形状的输入。

```
[batch_size, image_width, image_height, channels]
```

比如在数字分类任务中，池化层的输入有以下形状。

```
[batch_size, 28, 28, 20]
```

池化操作的输出有以下形状。

```
[batch_size, 14, 15, 20]
```

在本例中,卷积步骤的输出尺寸减少了 50%。这个步骤非常重要,因为它仅仅把重要的信息保留了下来,同时还减小了模型的复杂度,从而避免过拟合。

7.3.5 全连接层

在把一系列卷积和池化步骤堆叠起来后,就可以把它们和一个全连接层连接在一起,从输入图像提取的高层级的特征就会输入全连接层当中,此时就可以使用它们并基于这些特征来完成实际的分类,如图 7.9 所示。

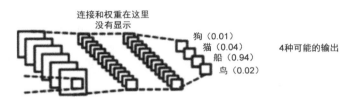

图 7.9 全连接层——每个节点都和相邻层的其他节点相连

例如,对于数字分类任务,可以在卷积和池化步骤之后接一个包含 1024 个神经元的全连接层以及一个 ReLU 激活函数来做实际的分类。这个全连接层接受的输入格式如下。

```
[batch_size, features]
```

所以,需要从 pool_layer1[①]重塑或者说压平输入特征映射以符合这个格式。可以用下面的代码重塑输出。

```
pool1_flat = tf.reshape(pool_layer1, [-1, 14 * 14 * 20])
```

在这个重塑函数中,用 −1 来表明批尺寸是动态决定的。在 pool_layer1 输出里,每一个样本的宽度都为 14,高度都为 14,同时有 20 个通道。

所以重塑操作的最后输出的形状如下。

```
[batch_size, 3136]
```

最后,可以用 TensorFlow 中的 dense() 函数通过指定神经元(单元)数量和最终激活

---

① 原文此处为 "pool_layer2",应为笔误。——译者注

函数来定义全连接层。

```
dense_layer = tf.layers.dense(inputs=pool1_flat, units=1024,
activation=tf.nn.relu)
```

取对数层

最后，我们需要一个取对数层，它接受全连接层的输出，然后产生原始的预测值。比如，在数字分类问题中，输出会是一个有 10 个值的张量，每个值代表 0~9 当中每个类的分数。如图 7.10 所示，对于数字分类示例定义一个取对数层，这里只需要 10 个输出，同时使用线性激活，这是 Tensorflow 中 dense()函数的默认设置。

```
logits_layer = tf.layers.dense(inputs=dense_layer, units=10)
```

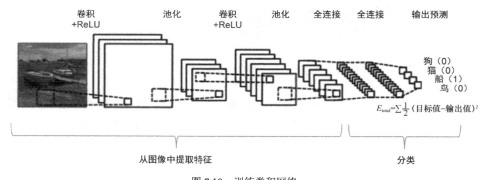

图 7.10 训练卷积网络

取对数层最后的输出是具有以下形状的张量。

```
[batch_size, 10]
```

正如前面所提到的，模型的取对数层会返回批次的原始预测。如图 7.11 所示，需要把这些数值转换为可以解释的格式。

- 输入样本 0~9 的预测类别。
- 每个可能类别的分数或者概率，比如，样本的概率是 0 或 1 等。

所以，这里预测的类别是 10 个概率值中最高的。可以像下面一样使用 argmax 函数获得这个值。

```
tf.argmax(input=logits_layer, axis=1)
```

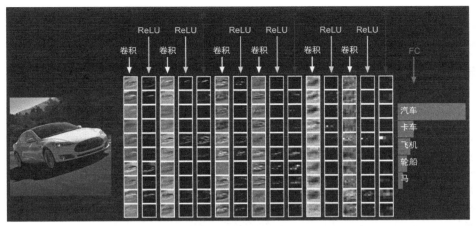

图 7.11　CNN 不同层的可视化（图片来源：GitHub 网站）

logits_layer 的形状如下。

```
[batch_size, 10]
```

因此，需要在预测当中找到最大值，即索引为 1 的维度。

最后，可以通过将 softmax 激活函数应用于 logits_layer 的输出来获得下一个值（这个值表示每个目标类的概率），这将把每个值压缩到 0～1。

```
tf.nn.softmax(logits_layer, name="softmax_tensor")
```

7.4　CNN 基础示例——MNIST 手写数字分类

本节将使用 MNIST 数据集完成一个实现 CNN 数字分类的完整示例。本节将构建一个包含两个卷积层和两个全连接层的简单模型。

首先，导入所需要的库。

```
%matplotlib inline
import matplotlib.pyplot as plt
import tensorflow as tf
import numpy as np
from sklearn.metrics import confusion_matrix
import math
```

然后，使用 TensorFlow 辅助函数来下载和预处理 MNIST 数据集。

```
from tensorflow.examples.tutorials.mnist import input_data
mnist_data = input_data.read_data_sets('data/MNIST/', one_hot=True)
```

输出如下。

```
Successfully downloaded train-images-idx3-ubyte.gz 9912422 bytes.
Extracting data/MNIST/train-images-idx3-ubyte.gz
Successfully downloaded train-labels-idx1-ubyte.gz 28881 bytes.
Extracting data/MNIST/train-labels-idx1-ubyte.gz
Successfully downloaded t10k-images-idx3-ubyte.gz 1648877 bytes.
Extracting data/MNIST/t10k-images-idx3-ubyte.gz
Successfully downloaded t10k-labels-idx1-ubyte.gz 4542 bytes.
Extracting data/MNIST/t10k-labels-idx1-ubyte.gz
```

数据集分为 3 个不相交的集合：训练集、验证集和测试集。通过以下代码输出每个集合中的图像数量。

```
print("- Number of images in the training
set:\t\t{}".format(len(mnist_data.train.labels)))
print("- Number of images in the test
set:\t\t{}".format(len(mnist_data.test.labels)))
print("- Number of images in the validation
set:\t{}".format(len(mnist_data.validation.labels)))

- Number of images in the training set: 55000
- Number of images in the test set: 10000
- Number of images in the validation set: 5000
```

因为图像真实的标签以独热（one-hot）编码的格式存储，所以会有一个包含 10 个数值的数组，除了图像类别的索引外，该数组其余位置皆为 0。为了方便后续使用，需要以整数的形式获得数据集的类别号。

```
mnist_data.test.cls_integer = np.argmax(mnist_data.test.labels, axis=1)
```

定义一些已知的变量，以便在后面的实现中使用。

```
# Default size for the input monocrome images of MNIST
image_size = 28

# Each image is stored as vector of this size.
image_size_flat = image_size * image_size
```

```
# The shape of each image
image_shape = (image_size, image_size)

# All the images in the mnist dataset are stored as a monocrome with only 1 channel
num_channels = 1

# Number of classes in the MNIST dataset from 0 till 9 which is 10
num_classes = 10
```

接下来，需要定义一个辅助函数来绘制数据集中的一些图像。该辅助函数将在9个子图的网格中绘制图像。

```
def plot_imgs(imgs, cls_actual, cls_predicted=None):
    assert len(imgs) == len(cls_actual) == 9
    # create a figure with 9 subplots to plot the images.
    fig, axes = plt.subplots(3, 3)
    fig.subplots_adjust(hspace=0.3, wspace=0.3)

    for i, ax in enumerate(axes.flat):
        # plot the image at the ith index
        ax.imshow(imgs[i].reshape(image_shape), cmap='binary')

        # labeling the images with the actual and predicted classes.
        if cls_predicted is None:
            xlabel = "True: {0}".format(cls_actual[i])
        else:
            xlabel = "True: {0}, Pred: {1}".format(cls_actual[i],cls_predicted[i])

        # Remove ticks from the plot.
        ax.set_xticks([])
        ax.set_yticks([])
        # Show the classes as the label on the x-axis.
        ax.set_xlabel(xlabel)
    plt.show()
```

现在从测试集中绘制一些图像并查看它的外观。

```
# Visualizing 9 images form the test set.
imgs = mnist_data.test.images[0:9]

# getting the actual classes of these 9 images
cls_actual = mnist_data.test.cls_integer[0:9]
```

```
#plotting the images
plot_imgs(imgs=imgs, cls_actual=cls_actual)
```

输出如图 7.12 所示。

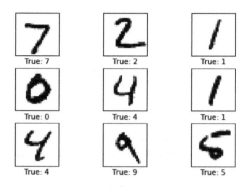

图 7.12　MNIST 数据集中部分样本的可视化

7.4.1　构建模型

下面开始搭建模型的核心。计算图包含本章前面提到的所有层。首先定义一些函数，这些函数将用于定义特定形状的变量并随机初始化它们。

```
def new_weights(shape):
    return tf.Variable(tf.truncated_normal(shape, stddev=0.05))

def new_biases(length):
    return tf.Variable(tf.constant(0.05, shape=[length]))
```

现在，根据输入层、输入通道、过滤器尺寸、过滤器数量以及是否使用池化参数，定义将负责创建新卷积层的函数。

```
def conv_layer(input, # the output of the previous layer.
               input_channels,
               filter_size,
               filters,
               use_pooling=True): # Use 2x2 max-pooling.

    # preparing the accepted shape of the input Tensor.
    shape = [filter_size, filter_size, input_channels, filters]

    # Create weights which means filters with the given shape.
```

7.4 CNN 基础示例——MNIST 手写数字分类

```
    filters_weights = new_weights(shape=shape)

    # Create new biases, one for each filter.
    filters_biases = new_biases(length=filters)

    # Calling the conve2d function as we explained above, were the strides parameter
    # has four values the first one for the image number and the last 1 for the
input image channel
    # the middle ones represents how many pixels the filter should move with in
the x and y axis
    conv_layer = tf.nn.conv2d(input=input,
                       filter=filters_weights,
                       strides=[1, 1, 1, 1],
                       padding='SAME')

    # Adding the biase to the output of the conv_layer.
    conv_layer += filters_biases

    # Use pooling to down-sample the image resolution?
    if use_pooling:

        # reduce the output feature map by max_pool layer
        pool_layer = tf.nn.max_pool(value=conv_layer,
                              ksize=[1, 2, 2, 1],
                              strides=[1, 2, 2, 1],
                              padding='SAME')

    # feeding the output to a ReLU activation function.
    relu_layer = tf.nn.relu(pool_layer)

    # return the final results after applying relu and the filter weights
    return relu_layer, filters_weights
```

正如前面提到的,池化层产生了一个 4 维张量。需要将这个 4 维张量压成 2 维张量以输入全连接层中。

```
def flatten_layer(layer):
    # Get the shape of layer.
    shape = layer.get_shape()

    # We need to flatten the layer which has the shape of The shape [num_images,
image_height, image_width, num_channels]
```

```
    # so that it has the shape of [batch_size, num_features] where
number_features is image_height * image_width * num_channels

    number_features = shape[1:4].num_elements()
    # Reshaping that to be fed to the fully connected layer
    flatten_layer = tf.reshape(layer, [-1, number_features])

    # Return both the flattened layer and the number of features.
    return flatten_layer, number_features
```

该函数创建一个全连接层，并假设输入是 2 维张量。

```
def fc_layer(input, # the flatten output.
             num_inputs, # Number of inputs from previous layer
             num_outputs, # Number of outputs
             use_relu=True): # Use ReLU on the output to remove negative
values

    # Creating the weights for the neurons of this fc_layer
    fc_weights = new_weights(shape=[num_inputs, num_outputs])
    fc_biases = new_biases(length=num_outputs)

    # Calculate the layer values by doing matrix multiplication of
    # the input values and fc_weights, and then add the fc_bias-values.
    fc_layer = tf.matmul(input, fc_weights) + fc_biases
    # if use RelU parameter is true
    if use_relu:
        relu_layer = tf.nn.relu(fc_layer)
        return relu_layer

    return fc_layer
```

在开始构建网络前，先给输入图像定义一个占位符，其中第一维用 None 来表示任意数量的图像。

```
input_values = tf.placeholder(tf.float32, shape=[None,image_size_flat],name=
'input_values')
```

正如前面提到的，卷积步骤要求输入图像是 4 维张量。所以，需要将输入图像重塑成以下形状。

```
[num_images, image_height, image_width, num_channels]
```

7.4 CNN 基础示例——MNIST 手写数字分类

重塑输入值以匹配这个格式。

```
input_image = tf.reshape(input_values, [-1, image_size, image_size, num_channels])
```

接下来，按照独热编码格式为实际的类别值定义另一个占位符。

```
y_actual = tf.placeholder(tf.float32, shape=[None, num_classes], name='y_actual')
```

同时，还需要定义一个占位符来存储实际类别的整数值。

```
y_actual_cls_integer = tf.argmax(y_actual, axis=1)
```

现在开始搭建第一个 CNN。

```
conv_layer_1, conv1_weights = \
        conv_layer(input=input_image,
                   input_channels=num_channels,
                   filter_size=filter_size_1,
                   filters=filters_1,
                   use_pooling=True)
```

检查第一个卷积层产生的输出张量的形状。

```
conv_layer_1
```

输出如下。

```
<tf.Tensor 'Relu:0' shape=(?, 14, 14, 16) dtype=float32>
```

接下来，构造第二个卷积层，并把第一个卷积层的输出输入其中。

```
conv_layer_2, conv2_weights = \
        conv_layer(input=conv_layer_1,
                   input_channels=filter1,
                   filter_size=filter_size_2,
                   filters=filters_2,
                   use_pooling=True)
```

同时还需要再次检查一下第二个卷积层的输出张量的形状。它的形状应该是(?, 7, 7, 36)，这里"?"表示任意数量的图像。

接下来，需要把 4 维张量压平以匹配全连接层所要求的格式，也就是要将它压平成一个 2 维张量。

```
flatten_layer, number_features = flatten_layer(conv_layer_2)
```

需要再次检查压平层中输出张量的形状。

```
flatten_layer
```

输出如下。

```
<tf.Tensor 'Reshape_1:0' shape=(?, 1764) dtype=float32>
```

接下来，构建全连接层，并把压平层的输出输入全连接层里。全连接层的输出在输入第二个全连接层之前还会输入 ReLU 激活函数里。

```
fc_layer_1 = fc_layer(input=flatten_layer,
                      num_inputs=number_features,
                      num_outputs=fc_num_neurons,
                      use_relu=True)
```

再次检查第一个全连接层中输出张量的形状。

```
fc_layer_1
```

输出如下。

```
<tf.Tensor 'Relu_2:0' shape=(?, 128) dtype=float32>
```

接下来，添加另一个全连接层，它将接受第一个全连接层的输出，然后对每幅图像产生一个长度为 10 的数组，该数组表示各个目标类别是正确类别的分数。

```
fc_layer_2 = fc_layer(input=fc_layer_1,
                      num_inputs=fc_num_neurons,
                      num_outputs=num_classes,
                      use_relu=False)
fc_layer_2
```

输出如下。

```
<tf.Tensor 'add_3:0' shape=(?, 10) dtype=float32>
```

7.4 CNN 基础示例——MNIST 手写数字分类

接下来，归一化第二个全连接层的分数并把它输入 softmax 激活函数中，它将数值压缩到 0~1。

```
y_predicted = tf.nn.softmax(fc_layer_2)
```

最后，用 TensorFlow 的 argmax 函数选择拥有最高概率的目标类别。

```
y_predicted_cls_integer = tf.argmax(y_predicted, axis=1)
```

1．成本函数

首先，需要定义性能指标，也就是交叉熵。如果预测的类别是正确的，交叉熵的数值就为 0。

```
cross_entropy = tf.nn.softmax_cross_entropy_with_logits(logits=fc_layer_2, labels=
                                                        y_actual)
```

接着，对上一步得到的交叉熵求均值来获得整个测试集上的单个性能指标。

```
model_cost = tf.reduce_mean(cross_entropy)
```

现在，因为需要优化或最小化损失函数，所以要使用 AdamOptimizer，它是类似于梯度下降法但更先进一些的优化方法。

```
model_optimizer =
tf.train.AdamOptimizer(learning_rate=1e-4).minimize(model_cost)
```

2．性能指标

为了展示输出，定义一个变量来判断预测类别是否等于真实类别。

```
model_correct_prediction = tf.equal(y_predicted_cls_integer, y_actual_cls_integer)
```

通过强制转换为布尔值，然后把正确分类的数量相加并求平均值来计算模型的准确率。

```
model_accuracy = tf.reduce_mean(tf.cast(model_correct_prediction,
tf.float32))
```

7.4.2 训练模型

现在创建一个会话变量来启动训练过程，该变量将负责执行之前定义的计算图。

```
session = tf.Session()
```

此外,需要初始化到目前为止已定义的变量。

```
session.run(tf.global_variables_initializer())
```

将分批提供图像以避免出现内存不足的错误。

```
train_batch_size = 64
```

在开始训练过程之前,需要定义一个辅助函数,该函数通过迭代训练批次来执行优化过程。

```
# number of optimization iterations performed so far
total_iterations = 0

def optimize(num_iterations):
    # Update globally the total number of iterations performed so far.

    global total_iterations

    for i in range(total_iterations,
                   total_iterations + num_iterations):

        # Generating a random batch for the training process
        # input_batch now contains a bunch of images from the training set and
        # y_actual_batch are the actual labels for the images in the input batch.
        input_batch, y_actual_batch = mnist_data.train.next_batch(train_batch_size)

        # Putting the previous values in a dict format for Tensorflow to automatically assign them to the input
        # placeholders that we defined above
        feed_dict = {input_values: input_batch,
                     y_actual: y_actual_batch}

        # Next up, we run the model optimizer on this batch of images
        session.run(model_optimizer, feed_dict=feed_dict)

        # Print the training status every 100 iterations.
        if i % 100 == 0:
            # measuring the accuracy over the training set.
```

```
            acc_training_set = session.run(model_accuracy,
feed_dict=feed_dict)
            #Printing the accuracy over the training set
            print("Iteration: {0:>6}, Accuracy Over the training set: {1:>6.1%}".
format(i + 1, acc_training_set))

    # Update the number of iterations performed so far
    total_iterations += num_iterations
```

另外定义一些辅助函数来帮助可视化模型的结果,并查看哪些图像被模型错误分类。

```
def plot_errors(cls_predicted, correct):
    # cls_predicted is an array of the predicted class number of each image in
the test set.

    # Extracting the incorrect images.
    incorrect = (correct == False)
    # Get the images from the test-set that have been
    # incorrectly classified.
    images = mnist_data.test.images[incorrect]
    # Get the predicted classes for those incorrect images.
    cls_pred = cls_predicted[incorrect]

    # Get the actual classes for those incorrect images.
    cls_true = mnist_data.test.cls_integer[incorrect]
    # Plot 9 of these images
    plot_imgs(imgs=imgs[0:9],
              cls_actual=cls_actual[0:9],
              cls_predicted=cls_predicted[0:9])
```

还可以绘制预测结果与真实类别的混淆矩阵。

```
def plot_confusionMatrix(cls_predicted):

 # cls_predicted is an array of the predicted class number of each image in the
test set.

 # Get the actual classes for the test-set.
 cls_actual = mnist_data.test.cls_integer

 # Generate the confusion matrix using sklearn.
 conf_matrix = confusion_matrix(y_true=cls_actual,
 y_pred=cls_predicted)
```

```python
# Print the matrix.
print(conf_matrix)

# visualizing the confusion matrix.
plt.matshow(conf_matrix)

plt.colorbar()
tick_marks = np.arange(num_classes)
plt.xticks(tick_marks, range(num_classes))
plt.yticks(tick_marks, range(num_classes))
plt.xlabel('Predicted class')
plt.ylabel('True class')

# Showing the plot
plt.show()
```

最后,定义一个辅助函数来辅助测量训练模型在测试集上的准确性。

```python
# measuring the accuracy of the trained model over the test set by
splitting it into small batches
test_batch_size = 256

def test_accuracy(show_errors=False,
                  show_confusionMatrix=False):

    #number of test images
    number_test = len(mnist_data.test.images)

    # define an array of zeros for the predicted classes of the test set which
    # will be measured in mini batches and stored it.
    cls_predicted = np.zeros(shape=number_test, dtype=np.int)

    # measuring the predicted classes for the testing batches.

    # Starting by the batch at index 0.
    i= 0

    while i < number_test:
        # The ending index for the next batch to be processed is j.
        j = min(i + test_batch_size, number_test)
```

7.4 CNN 基础示例——MNIST 手写数字分类

```
    # Getting all the images form the test set between the start and end indices
    input_images = mnist_data.test.images[i:j, :]

    # Get the acutal labels for those images.
    actual_labels = mnist_data.test.labels[i:j, :]

    # Create a feed-dict with the corresponding values for the input placeholder values
    feed_dict = {input_values: input_images,
                 y_actual: actual_labels}

    cls_predicted[i:j] = session.run(y_predicted_cls_integer, feed_dict=feed_dict)

    # Setting the start of the next batch to be the end of the one that we just processed j
    i= j

# Get the actual class numbers of the test images.
cls_actual = mnist_data.test.cls_integer

# Check if the model predictions are correct or not
correct = (cls_actual == cls_predicted)

# Summing up the correct examples
correct_number_images = correct.sum()

# measuring the accuracy by dividing the correclty classified ones with total number of images in the test set.
testset_accuracy = float(correct_number_images) / number_test

# showing the accuracy.
print("Accuracy on Test-Set: {0:.1%} ({1} / {2})".format(testset_accuracy, correct_number_images, number_test))

# showing some examples form the incorrect ones.
if show_errors:
    print("Example errors:")
    plot_errors(cls_predicted=cls_predicted, correct=correct)

# Showing the confusion matrix of the test set predictions
if show_confusionMatrix:
```

```
print("Confusion Matrix:")
plot_confusionMatrix(cls_predicted=cls_predicted)
```

在进行优化之前，在测试集上输出已创建模型的准确率。

```
test_accuracy()
```

输出如下。

```
Accuracy on Test-Set: 4.1% (410 / 10000)
```

通过运行一次迭代的优化过程，可以了解到优化过程实际上增强了模型功能，以便将图像划分到正确的类中。

```
optimize(num_iterations=1)
```

输出如下。

```
Iteration: 1, Accuracy Over the training set: 4.7%
test_accuracy()
```

输出如下。

```
Accuracy on Test-Set: 4.4% (437 / 10000)
```

现在开始训练，即开始10000次迭代的长时间优化过程。

```
optimize(num_iterations=9999) #We have already performed 1 iteration.
```

在输出结束时，我们应该得到非常接近以下输出的内容。

```
Iteration: 7301, Accuracy Over the training set: 96.9%
Iteration: 7401, Accuracy Over the training set: 100.0%
Iteration: 7501, Accuracy Over the training set: 98.4%
Iteration: 7601, Accuracy Over the training set: 98.4%
Iteration: 7701, Accuracy Over the training set: 96.9%
Iteration: 7801, Accuracy Over the training set: 96.9%
Iteration: 7901, Accuracy Over the training set: 100.0%
Iteration: 8001, Accuracy Over the training set: 98.4%
Iteration: 8101, Accuracy Over the training set: 96.9%
Iteration: 8201, Accuracy Over the training set: 100.0%
Iteration: 8301, Accuracy Over the training set: 98.4%
```

```
Iteration: 8401, Accuracy Over the training set: 98.4%
Iteration: 8501, Accuracy Over the training set: 96.9%
Iteration: 8601, Accuracy Over the training set: 100.0%
Iteration: 8701, Accuracy Over the training set: 98.4%
Iteration: 8801, Accuracy Over the training set: 100.0%
Iteration: 8901, Accuracy Over the training set: 98.4%
Iteration: 9001, Accuracy Over the training set: 100.0%
Iteration: 9101, Accuracy Over the training set: 96.9%
Iteration: 9201, Accuracy Over the training set: 98.4%
Iteration: 9301, Accuracy Over the training set: 98.4%
Iteration: 9401, Accuracy Over the training set: 100.0%
Iteration: 9501, Accuracy Over the training set: 100.0%
Iteration: 9601, Accuracy Over the training set: 98.4%
Iteration: 9701, Accuracy Over the training set: 100.0%
Iteration: 9801, Accuracy Over the training set: 100.0%
Iteration: 9901, Accuracy Over the training set: 100.0%
Iteration: 10001, Accuracy Over the training set: 98.4%
```

现在，查看模型在测试集上如何泛化。

```
test_accuracy(show_errors=True,
              show_confusionMatrix=True)
```

输出如下。

```
Accuracy on Test-Set: 92.8% (9281 / 10000)
Example errors:
```

输出结果如图 7.13 所示。

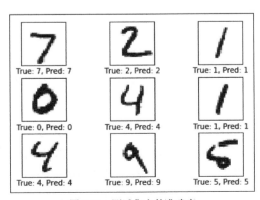

图 7.13　测试集上的准确率

对应的混淆矩阵如下。

```
[[ 971    0    2    2    0    4    0    1    0    0]
 [   0 1110    4    2    1    2    3    0   13    0]
 [  12    2  949   15   16    3    4   17   14    0]
 [   5    3   14  932    0   34    0   13    6    3]
 [   1    2    3    0  931    1    8    2    3   31]
 [  12    1    4   13    3  852    2    1    3    1]
 [  21    4    5    2   18   34  871    1    2    0]
 [   1   10   26    5    5    0    0  943    2   36]
 [  16    5   10   27   16   48    5   13  815   19]
 [  12    5    5   11   38   10    0   18    3  907]]
```

输出如图 7.14 所示。

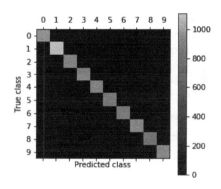

图 7.14 测试集上的混淆矩阵

有趣的是，在使用基本卷积网络时，实际上在测试中获得了接近 93%的准确率。这个实现和结果展示了一个简单的卷积网络可以做什么。

7.5 总结

本章介绍了 CNN 的工作原理和技术细节，同时还介绍了如何在 TensorFlow 中实现 CNN 的基本架构。

下一章将演示更高级的体系结构，这些体系结构可用于数据科学家广泛使用的图像数据集中的目标检测。我们还将看到 CNN 的优点以及它如何模仿人类对物体的理解：首先实现对象的基本特征，然后在它们上构建更高级的语义特征来实现分类。虽然这个过程会在我们的脑海中"一闪而过"，但它确实是我们识别物体时实际发生的事情。

第 8 章
目标检测——CIFAR-10 示例

在介绍了**卷积神经网络**（CNN）背后的基础知识和直觉（动机）之后，本章将在物体检测方面最著名的数据集上进行演示。同样地，你将会看到 CNN 前面的一些层是如何提取关于物体的一些基本特征的，而最后的卷积层将会提取更多的语义级特征，这些特征都是从前面层的基本特征中构建而来的。

本章将包含以下两个主题。

- 目标检测。
- CIFAR-10 图像目标检测——构建与训练模型。

8.1 目标检测

维基百科指出：

"目标检测——计算机视觉领域的技术，旨在查找和识别位于图像或视频序列中的对象。尽管当平移或旋转物体时，物体的图像可能有不同的视点、不同的大小和尺度，但是人类都能够很容易地识别出图像中的多个物体。甚至当物体被部分遮挡时，人类都能够识别出这些物体。但这项任务对于计算机视觉系统来说仍然是一项挑战。几十年来，完成这项任务的很多方法都已经实现了。"

图像分析是深度学习中最突出的领域之一。图像易于生成和处理，它们也是机器学习中的正确数据类型：对于人类容易理解，但对于计算机很难。所以图像分析在深度神经网络历史中发挥了关键作用也是不足为奇的。

随着自动驾驶汽车、面部检测、智能视频监控和人口统计解决方案的兴起，迫切需要快速、准确的目标检测系统。这些系统不仅包括图像中的目标识别和分类，还可以通过在它们周围绘制适当的方框来定位每一个目标。这也使得目标检测比传统的计算机视觉前身——图像分类任务更加困难。

本章的重点是目标检测，即找到图像中有哪些物体。例如，如图 8.1 所示，一辆自动驾驶汽车需要检测道路上的其他汽车。目标检测有很多复杂的算法，它们通常需要庞大的数据集、非常深的卷积网络和很长的训练时间。

图 8.1　目标检测的例子（图片来源：B. C. Russell, A. Torralba, C. Liu, R. Fergus, W. T. Freeman, Object Detection by Scene Alignment, Advances in Neural Information Processing Systems, 2007）

8.2　CIFAR-10 目标图像检测——构建与训练模型

接下来的例子将展示如何建立 CNN 来对 CIFAR-10 数据集中的图像进行分类。本节将使用一个带有几个卷积和全连接层的简单卷积神经网络。

尽管网络架构非常简单，但是我们可以看到它在尝试检测 CIFAR-10 数据集图像中的物体时表现得还是非常好的。

下面来看具体实现。

8.2.1　使用软件包

首先，导入所有具体实现需要用到的软件包。

```
%matplotlib inline
%config InlineBackend.figure_format = 'retina'
```

```python
from urllib.request import urlretrieve
from os.path import isfile, isdir
from tqdm import tqdm
import tarfile
import numpy as np
import random
import matplotlib.pyplot as plt
from sklearn.preprocessing import LabelBinarizer
from sklearn.preprocessing import OneHotEncoder

import pickle
import tensorflow as tf
```

8.2.2 加载 CIFAR-10 数据集

在对 CIFAR-10 示例的代码实现中,我们使用了 CIFAR-10 数据集,它是目标检测中使用最广泛的数据集之一。如果尚未下载该数据集,可以先定义一个辅助类来下载和提取 CIFAR-10 数据集。

```python
cifar10_batches_dir_path = 'cifar-10-batches-py'

tar_gz_filename = 'cifar-10-python.tar.gz'

class DLProgress(tqdm):
    last_block = 0

    def hook(self, block_num=1, block_size=1, total_size=None):
        self.total = total_size
        self.update((block_num - self.last_block) * block_size)
        self.last_block = block_num

if not isfile(tar_gz_filename):
    with DLProgress(unit='B', unit_scale=True, miniters=1, desc='CIFAR-10 Python Images Batches') as pbar:
        urlretrieve(
            'https://www.cs.toronto.edu/~kriz/cifar-10-python.tar.gz',
            tar_gz_filename,
            pbar.hook)

if not isdir(cifar10_batches_dir_path):
    with tarfile.open(tar_gz_filename) as tar:
```

```
tar.extractall()
tar.close()
```

在下载和提取 CIFAR-10 数据集之后，可以发现它已经分为了 5 批。CIFAR-10 包含如下 10 个类别的图像。

- 飞机（airplane）
- 汽车（automobile）
- 鸟（bird）
- 猫（cat）
- 鹿（deer）
- 狗（dog）
- 青蛙（frog）
- 马（horse）
- 船（ship）
- 卡车（truck）

在深入构建网络的核心之前，先做一些数据分析和预处理。

8.2.3 数据分析与预处理

这里需要分析数据集并进行一些基本的预处理。首先，定义一些辅助函数，它能够从现有的 5 个批次中加载特定批次，并输出有关批次及其样本的一些分析情况。

```
# Defining a helper function for loading a batch of images
def load_batch(cifar10_dataset_dir_path, batch_num):
    with open(cifar10_dataset_dir_path + '/data_batch_' + str(batch_num), mode='rb') as file:
        batch = pickle.load(file, encoding='latin1')
    input_features = batch['data'].reshape((len(batch['data']), 3, 32, 32)).transpose(0, 2, 3, 1)
    target_labels = batch['labels']

    return input_features, target_labels
```

然后，定义一个函数，它能够帮助我们显示特定批次中特定样本的统计数据。

```
#Defining a function to show the stats for batch ans specific sample
def batch_image_stats(cifar10_dataset_dir_path, batch_num, sample_num):

    batch_nums = list(range(1, 6))

    #checking if the batch_num is a valid batch number
    if batch_num not in batch_nums:
        print('Batch Num is out of Range. You can choose from these Batch nums: {}'.format(batch_nums))
        return None

    input_features, target_labels = load_batch(cifar10_dataset_dir_path,batch_num)

    #checking if the sample_num is a valid sample number
    if not (0 <= sample_num < len(input_features)):
        print('{} samples in batch {}. {} is not a valid sample number.'.format(len(input_features), batch_num, sample_num))
        return None

    print('\nStatistics of batch number {}:'.format(batch_num))
    print('Number of samples in this batch: {}'.format(len(input_features)))
    print('Per class counts of each Label: {}'.format(dict(zip(*np.unique(target_labels, return_counts=True)))))

    image = input_features[sample_num]
    label = target_labels[sample_num]
    cifar10_class_names = ['airplane', 'automobile', 'bird', 'cat', 'deer', 'dog', 'frog', 'horse', 'ship', 'truck']

    print('\nSample Image Number {}:'.format(sample_num))
    print('Sample image - Minimum pixel value: {} Maximum pixel value: {}'.format(image.min(), image.max()))
    print('Sample image - Shape: {}'.format(image.shape))
    print('Sample Label - Label Id: {} Name: {}'.format(label, cifar10_class_names[label]))
    plt.axis('off')
    plt.imshow(image)
```

现在，可以使用此函数来处理数据集并可视化特定图像。

```
# Explore a specific batch and sample from the dataset
batch_num = 3
sample_num = 6
batch_image_stats(cifar10_batches_dir_path, batch_num, sample_num)
```

输出结果如下所示。

```
Statistics of batch number 3:
Number of samples in this batch: 10000
Per class counts of each Label: {0: 994, 1: 1042, 2: 965, 3: 997, 4: 990,
5: 1029, 6: 978, 7: 1015, 8: 961, 9: 1029}

Sample Image Number 6:
Sample image - Minimum pixel value: 30 Maximum pixel value: 242
Sample image - Shape: (32, 32, 3)
Sample Label - Label Id: 8 Name: ship
```

输出的图像如图 8.2 所示。

在继续将数据集提供给模型之前，还需要将它归一化到 0～1。

批量标准化优化了网络训练。已证明它有以下几个好处。

图 8.2　批次 3 中的第 6 张样本图像

- **更快的训练速度**。由于在网络的前向传递期间存在额外的计算，并且在反向传播过程中有待训练的额外超参数，因此训练的每一步将会变得更慢。但是，它能够更快速地收敛，因此整体训练速度将会更快。
- **更大的学习率**。通常来说，梯度下降算法需要较小的学习率才能使得网络收敛到损失函数的最小值。随着神经网络越来越深，它们的梯度值在反向传播过程中变得越来越小，因此通常需要更多的迭代次数。批量标准化的想法允许我们使用更大的学习率，这进一步提高了网络的训练速度。
- **易于初始化权重**。权重初始化可能很困难，当使用深度神经网络时权重初始化会更加困难。批量归一化使得我们在选择初始化权重时可以不用那么讲究。

接下来，定义一个函数，该函数负责归一化一系列输入图像，使得这些图像的像素值都处于 0～1。

```
#Normalize CIFAR-10 images to be in the range of [0,1]

def normalize_images(images):
    # initial zero ndarray
    normalized_images = np.zeros_like(images.astype(float))
    # The first images index is number of images where the other indices indicates
    # hieight, width and depth of the image
    num_images = images.shape[0]
    # Computing the minimum and maximum value of the input image to do the normalization based on them
    maximum_value, minimum_value = images.max(), images.min()
    # Normalize all the pixel values of the images to be from 0 to 1
    for img in range(num_images):
        normalized_images[img,...] = (images[img, ...] - float(minimum_value)) / float(maximum_value - minimum_value)

    return normalized_images
```

接下来，还需要实现另一个辅助函数来编码输入图像的标签。该函数将会用到 sklearn 中的独热编码，其中每个图像标签由向量表示，该向量中只有代表图像类别索引的元素为 1，其余元素全为 0。

输出向量的大小取决于数据集中的物体类别数目，对于 CIFAR-10 数据来说，这就是 10 个类别。

```
#encoding the input images. Each image will be represented by a vector of zeros except for the class index of the image
# that this vector represents. The length of this vector depends on number of classes that we have
# the dataset which is 10 in CIFAR-10

def one_hot_encode(images):
    num_classes = 10
    #use sklearn helper function of OneHotEncoder() to do that
    encoder = OneHotEncoder(num_classes)
    #resize the input images to be 2D
    input_images_resized_to_2d = np.array(images).reshape(-1,1)
    one_hot_encoded_targets = encoder.fit_transform(input_images_resized_to_2d)
    return one_hot_encoded_targets.toarray()
```

下面调用前面的辅助函数来进行预处理,并要保留数据集,以便之后还可以使用它。

```
def preprocess_persist_data(cifar10_batches_dir_path, normalize_images,
one_hot_encode):
    num_batches = 5
    valid_input_features = []
    valid_target_labels = []

    for batch_ind in range(1, num_batches + 1):
        #Loading batch
        input_features, target_labels =
load_batch(cifar10_batches_dir_path, batch_ind)
        num_validation_images = int(len(input_features) * 0.1)

        # Preprocess the current batch and perisist it for future use
        input_features = normalize_images(input_features[:-
num_validation_images])
        target_labels = one_hot_encode( target_labels[:-
num_validation_images])
        #Persisting the preprocessed batch
        pickle.dump((input_features, target_labels),
open('preprocess_train_batch_' + str(batch_ind) + '.p', 'wb'))

        # Define a subset of the training images to be used for validating
our model
        valid_input_features.extend(input_features[-
num_validation_images:])
        valid_target_labels.extend(target_labels[-num_validation_images:])

    # Preprocessing and persisting the validationi subset
    input_features = normalize_images( np.array(valid_input_features))
    target_labels = one_hot_encode(np.array(valid_target_labels))
    pickle.dump((input_features, target_labels), open('preprocess_valid.p',
'wb'))

    #Now it's time to preporcess and persist the test batche
    with open(cifar10_batches_dir_path + '/test_batch', mode='rb') as file:
        test_batch = pickle.load(file, encoding='latin1')

    test_input_features =
test_batch['data'].reshape((len(test_batch['data']), 3, 32,
32)).transpose(0, 2, 3, 1)
    test_input_labels = test_batch['labels']
```

```
    # Normalizing and encoding the test batch
    input_features = normalize_images( np.array(test_input_features))
    target_labels = one_hot_encode(np.array(test_input_labels))
    pickle.dump((input_features, target_labels), open('preprocess_test.p',
'wb'))
# Calling the helper function above to preprocess and persist the training,
validation, and testing set
preprocess_persist_data(cifar10_batches_dir_path, normalize_images,
one_hot_encode)
```

现在,将预处理数据存入磁盘中。

还需要在训练过程的不同时期加载用于运行训练模型的验证集。

```
# Load the Preprocessed Validation data
valid_input_features, valid_input_labels =
pickle.load(open('preprocess_valid.p', mode='rb'))
```

8.2.4 建立网络

下面建立分类应用的核心部分——CNN 架构中的计算图。为了最大化这种实现的好处,这里不会直接使用 TensorFlow 中对应层的 API。相反,本次实现将会用到它的 TensorFlow 神经网络版本。

首先,定义模型输入占位符,它将输入图像、目标类和 droupout 层中的保留概率参数(dropout 层有助于我们丢弃一些连接来降低架构的复杂性,从而降低过拟合的概率)。

```
# Defining the model inputs
def images_input(img_shape):
 return tf.placeholder(tf.float32, (None, ) + img_shape,
name="input_images")

def target_input(num_classes):

 target_input = tf.placeholder(tf.int32, (None, num_classes),
name="input_images_target")
 return target_input

#define a function for the dropout layer keep probability
```

```python
def keep_prob_input():
 return tf.placeholder(tf.float32, name="keep_prob")
```

接下来,需要使用TensorFlow神经网络实现版本来构建后接最大池化操作的卷积层。

```python
# Applying a convolution operation to the input tensor followed by max
pooling
def conv2d_layer(input_tensor, conv_layer_num_outputs, conv_kernel_size,
conv_layer_strides, pool_kernel_size, pool_layer_strides):

 input_depth = input_tensor.get_shape()[3].value
 weight_shape = conv_kernel_size + (input_depth, conv_layer_num_outputs,)

 #Defining layer weights and biases
 weights = tf.Variable(tf.random_normal(weight_shape))
 biases = tf.Variable(tf.random_normal((conv_layer_num_outputs,)))

 #Considering the biase variable
 conv_strides = (1,) + conv_layer_strides + (1,)

 conv_layer = tf.nn.conv2d(input_tensor, weights, strides=conv_strides, padding='SAME')
 conv_layer = tf.nn.bias_add(conv_layer, biases)

 conv_kernel_size = (1,) + conv_kernel_size + (1,)

 pool_strides = (1,) + pool_layer_strides + (1,)
 pool_layer = tf.nn.max_pool(conv_layer, ksize=conv_kernel_size, strides=pool_strides, padding='SAME')
 return pool_layer
```

如前文所述,最大池化操作的输出是一个4维张量,这与全连接层所需要的输入格式不兼容。因此,还需要实现一个展平层来将最大池化层的输出从4维张量转换为2维张量。

```python
#Flatten the output of max pooling layer to be fing to the fully connected
layer which only accepts the output
# to be in 2D
def flatten_layer(input_tensor):
 return tf.contrib.layers.flatten(input_tensor)
```

接下来,需要定义一个辅助函数,该函数能够在之前的架构中添加一个全连接层。

```python
#Define the fully connected layer that will use the flattened output of the
stacked convolution layers
#to do the actuall classification
def fully_connected_layer(input_tensor, num_outputs):
    return tf.layers.dense(input_tensor, num_outputs)
```

最后,在使用这些辅助函数来建立整个网络结构时,还需要创建另一个函数来接受全连接层的输出,并生成与数据集中所具有的类别数目相对应的 10 个实值。

```python
#Defining the output function
def output_layer(input_tensor, num_outputs):
    return tf.layers.dense(input_tensor, num_outputs)
```

因此,继续定义一个函数,该函数可以将所有这些技术细节融合在一起,并创建一个具有 3 个卷积层的 CNN。其中每一个卷积层后面都有一个最大池化操作。同样,也会有两个全连接层,其中每一个全连接层后面都有一个 droupout 层,droupout 层用于降低模型复杂性和防止过拟合。最后,输出层将会生成包含 10 个实值的向量,其中每一个值代表每一个样本正确分类的得分。

```python
def build_convolution_net(image_data, keep_prob):
    
    # Applying 3 convolution layers followed by max pooling layers
    conv_layer_1 = conv2d_layer(image_data, 32, (3,3), (1,1), (3,3), (3,3))
    conv_layer_2 = conv2d_layer(conv_layer_1, 64, (3,3), (1,1), (3,3), (3,3))
    conv_layer_3 = conv2d_layer(conv_layer_2, 128, (3,3), (1,1), (3,3), (3,3))
    
    # Flatten the output from 4D to 2D to be fed to the fully connected layer
    flatten_output = flatten_layer(conv_layer_3)
    
    # Applying 2 fully connected layers with drop out
    fully_connected_layer_1 = fully_connected_layer(flatten_output, 64)
    fully_connected_layer_1 = tf.nn.dropout(fully_connected_layer_1, keep_prob)
    fully_connected_layer_2 = fully_connected_layer(fully_connected_layer_1, 32)
    fully_connected_layer_2 = tf.nn.dropout(fully_connected_layer_2, keep_prob)
    
    #Applying the output layer while the output size will be the number of
    categories that we have
    #in CIFAR-10 dataset
    output_logits = output_layer(fully_connected_layer_2, 10)
    
    #returning output
    return output_logits
```

现在调用前面的辅助函数来建立网络，并定义其损失和优化标准。

```
#Using the helper function above to build the network

#First off, let's remove all the previous inputs, weights, biases form the
previous runs
tf.reset_default_graph()

# Defining the input placeholders to the convolution neural network
input_images = images_input((32, 32, 3))
input_images_target = target_input(10)
keep_prob = keep_prob_input()

# Building the models
logits_values = build_convolution_net(input_images, keep_prob)

# Name logits Tensor, so that is can be loaded from disk after training
logits_values = tf.identity(logits_values, name='logits')

# defining the model loss
model_cost =
tf.reduce_mean(tf.nn.softmax_cross_entropy_with_logits(logits=logits_values,
labels=input_images_target))

# Defining the model optimizer
model_optimizer = tf.train.AdamOptimizer().minimize(model_cost)

# Calculating and averaging the model accuracy
correct_prediction = tf.equal(tf.argmax(logits_values, 1),
tf.argmax(input_images_target, 1))
accuracy = tf.reduce_mean(tf.cast(correct_prediction, tf.float32),
name='model_accuracy')
tests.test_conv_net(build_convolution_net)
```

既然网络的计算架构已经建立起来了，下面就开始训练过程并查看结果。

8.2.5 训练模型

首先，定义一个辅助函数，以便开始训练过程。该函数以输入图像、目标类别的独热编码和 droupout 中的保留概率值作为输入。然后，它将这些值输入计算图并调用模型优化器。

```
#Define a helper function for kicking off the training process
def train(session, model_optimizer, keep_probability, in_feature_batch,
target_batch):
```

8.2 CIFAR-10 目标图像检测——构建与训练模型

```
session.run(model_optimizer, feed_dict={input_images: in_feature_batch,
input_images_target: target_batch, keep_prob: keep_probability})
```

因为需要在训练过程的不同时间步长验证模型,所以需要定义一个辅助函数来输出模型在验证集上的准确率。

```
#Defining a helper funcitno for print information about the model accuracy
and it's validation accuracy as well
def print_model_stats(session, input_feature_batch, target_label_batch,
model_cost, model_accuracy):
    validation_loss = session.run(model_cost, feed_dict={input_images:
input_feature_batch, input_images_target: target_label_batch, keep_prob:1.0})
    validation_accuracy = session.run(model_accuracy,
feed_dict={input_images: input_feature_batch, input_images_target:
target_label_batch, keep_prob: 1.0})
    print("Valid Loss: %f" %(validation_loss))
    print("Valid accuracy: %f" % (validation_accuracy))
```

还需要定义模型超参数,可以使用它来微调模型,以获得更好的性能。

```
# Model Hyperparameters
num_epochs = 100
batch_size = 128
keep_probability = 0.5
```

现在,可以开始训练过程了,但是只针对 CIFAR-10 数据集中的单个批次,还要查看基于该批次数据的模型准确率。

然而,在此之前,需要定义一个辅助函数,该函数将加载一批训练数据,并将输入图像和目标类别分开。

```
# Splitting the dataset features and labels to batches
def batch_split_features_labels(input_features, target_labels,
train_batch_size):
    for start in range(0, len(input_features), train_batch_size):
        end = min(start + train_batch_size, len(input_features))
        yield input_features[start:end], target_labels[start:end]

#Loading the persisted preprocessed training batches
def load_preprocess_training_batch(batch_id, batch_size):
    filename = 'preprocess_train_batch_' + str(batch_id) + '.p'
    input_features, target_labels = pickle.load(open(filename, mode='rb'))
```

```
    # Returning the training images in batches according to the batch size
defined above
    return batch_split_features_labels(input_features, target_labels,
train_batch_size)
```

现在正式开始针对一个批次的训练过程。

```
print('Training on only a Single Batch from the CIFAR-10 Dataset...')
with tf.Session() as sess:

    # Initializing the variables
    sess.run(tf.global_variables_initializer())

    # Training cycle
    for epoch in range(num_epochs):
    batch_ind = 1

    for batch_features, batch_labels in
load_preprocess_training_batch(batch_ind, batch_size):
    train(sess, model_optimizer, keep_probability, batch_features,
batch_labels)

    print('Epoch number {:>2}, CIFAR-10 Batch Number {}: '.format(epoch + 1,
batch_ind), end='')
    print_model_stats(sess, batch_features, batch_labels, model_cost,
accuracy)
```

输出如下。

```
.
.
.
Epoch number 85, CIFAR-10 Batch Number 1: Valid Loss: 1.490792
Valid accuracy: 0.550000
Epoch number 86, CIFAR-10 Batch Number 1: Valid Loss: 1.487118
Valid accuracy: 0.525000
Epoch number 87, CIFAR-10 Batch Number 1: Valid Loss: 1.309082
Valid accuracy: 0.575000
Epoch number 88, CIFAR-10 Batch Number 1: Valid Loss: 1.446488
Valid accuracy: 0.475000
Epoch number 89, CIFAR-10 Batch Number 1: Valid Loss: 1.430939
```

```
Valid accuracy: 0.550000
Epoch number 90, CIFAR-10 Batch Number 1: Valid Loss: 1.484480
Valid accuracy: 0.525000
Epoch number 91, CIFAR-10 Batch Number 1: Valid Loss: 1.345774
Valid accuracy: 0.575000
Epoch number 92, CIFAR-10 Batch Number 1: Valid Loss: 1.425942
Valid accuracy: 0.575000

Epoch number 93, CIFAR-10 Batch Number 1: Valid Loss: 1.451115
Valid accuracy: 0.550000
Epoch number 94, CIFAR-10 Batch Number 1: Valid Loss: 1.368719
Valid accuracy: 0.600000
Epoch number 95, CIFAR-10 Batch Number 1: Valid Loss: 1.336483
Valid accuracy: 0.600000
Epoch number 96, CIFAR-10 Batch Number 1: Valid Loss: 1.383425
Valid accuracy: 0.575000
Epoch number 97, CIFAR-10 Batch Number 1: Valid Loss: 1.378877
Valid accuracy: 0.625000
Epoch number 98, CIFAR-10 Batch Number 1: Valid Loss: 1.343391
Valid accuracy: 0.600000
Epoch number 99, CIFAR-10 Batch Number 1: Valid Loss: 1.319342
Valid accuracy: 0.625000
Epoch number 100, CIFAR-10 Batch Number 1: Valid Loss: 1.340849
Valid accuracy: 0.525000
```

可以看到，当仅在单个批次上进行训练时，经验证得到的准确率并不高。下面来看基于模型的完整训练过程，经验证得到的准确率将会如何变化。

```
model_save_path = './cifar-10_classification'

with tf.Session() as sess:
  # Initializing the variables
  sess.run(tf.global_variables_initializer())

  # Training cycle
  for epoch in range(num_epochs):

    # iterate through the batches
    num_batches = 5

    for batch_ind in range(1, num_batches + 1):
    for batch_features, batch_labels in
    load_preprocess_training_batch(batch_ind, batch_size):
```

```python
        train(sess, model_optimizer, keep_probability, batch_features,
batch_labels)

        print('Epoch number{:>2}, CIFAR-10 Batch Number {}: '.format(epoch + 1,
batch_ind), end='')
        print_model_stats(sess, batch_features, batch_labels, model_cost,
accuracy)

# Save the trained Model
saver = tf.train.Saver()
save_path = saver.save(sess, model_save_path)
```

输出如下。

```
.
.
.
Epoch number94, CIFAR-10 Batch Number 5: Valid Loss: 0.316593
Valid accuracy: 0.925000
Epoch number95, CIFAR-10 Batch Number 1: Valid Loss: 0.285429
Valid accuracy: 0.925000
Epoch number95, CIFAR-10 Batch Number 2: Valid Loss: 0.347411
Valid accuracy: 0.825000
Epoch number95, CIFAR-10 Batch Number 3: Valid Loss: 0.232483
Valid accuracy: 0.950000
Epoch number95, CIFAR-10 Batch Number 4: Valid Loss: 0.294707
Valid accuracy: 0.900000
Epoch number95, CIFAR-10 Batch Number 5: Valid Loss: 0.299490
Valid accuracy: 0.975000
Epoch number96, CIFAR-10 Batch Number 1: Valid Loss: 0.302191
Valid accuracy: 0.950000
Epoch number96, CIFAR-10 Batch Number 2: Valid Loss: 0.347043
Valid accuracy: 0.750000
Epoch number96, CIFAR-10 Batch Number 3: Valid Loss: 0.252851
Valid accuracy: 0.875000
Epoch number96, CIFAR-10 Batch Number 4: Valid Loss: 0.291433
Valid accuracy: 0.950000
Epoch number96, CIFAR-10 Batch Number 5: Valid Loss: 0.286192
Valid accuracy: 0.950000
Epoch number97, CIFAR-10 Batch Number 1: Valid Loss: 0.277105
Valid accuracy: 0.950000
Epoch number97, CIFAR-10 Batch Number 2: Valid Loss: 0.305842
Valid accuracy: 0.850000
```

```
Epoch number97, CIFAR-10 Batch Number 3: Valid Loss: 0.215272
Valid accuracy: 0.950000
Epoch number97, CIFAR-10 Batch Number 4: Valid Loss: 0.313761
Valid accuracy: 0.925000
Epoch number97, CIFAR-10 Batch Number 5: Valid Loss: 0.313503
Valid accuracy: 0.925000
Epoch number98, CIFAR-10 Batch Number 1: Valid Loss: 0.265828
Valid accuracy: 0.925000
Epoch number98, CIFAR-10 Batch Number 2: Valid Loss: 0.308948
Valid accuracy: 0.800000
Epoch number98, CIFAR-10 Batch Number 3: Valid Loss: 0.232083
Valid accuracy: 0.950000
Epoch number98, CIFAR-10 Batch Number 4: Valid Loss: 0.298826
Valid accuracy: 0.925000
Epoch number98, CIFAR-10 Batch Number 5: Valid Loss: 0.297230
Valid accuracy: 0.950000
Epoch number99, CIFAR-10 Batch Number 1: Valid Loss: 0.304203
Valid accuracy: 0.900000
Epoch number99, CIFAR-10 Batch Number 2: Valid Loss: 0.308775
Valid accuracy: 0.825000
Epoch number99, CIFAR-10 Batch Number 3: Valid Loss: 0.225072
Valid accuracy: 0.925000
Epoch number99, CIFAR-10 Batch Number 4: Valid Loss: 0.263737
Valid accuracy: 0.925000
Epoch number99, CIFAR-10 Batch Number 5: Valid Loss: 0.278601
Valid accuracy: 0.950000
Epoch number100, CIFAR-10 Batch Number 1: Valid Loss: 0.293509
Valid accuracy: 0.950000
Epoch number100, CIFAR-10 Batch Number 2: Valid Loss: 0.303817
Valid accuracy: 0.875000
Epoch number100, CIFAR-10 Batch Number 3: Valid Loss: 0.244428
Valid accuracy: 0.900000
Epoch number100, CIFAR-10 Batch Number 4: Valid Loss: 0.280712
Valid accuracy: 0.925000
Epoch number100, CIFAR-10 Batch Number 5: Valid Loss: 0.278625
Valid accuracy: 0.950000
```

8.2.6 测试模型

现在根据 CIFAR-10 数据集中的测试集部分来测试训练好的模型。首先，定义一个辅助函数，该函数有助于可视化一些样本图像及其对应的真实标签。

```python
#A helper function to visualize some samples and their corresponding
predictions
def display_samples_predictions(input_features, target_labels,
samples_predictions):

    num_classes = 10

    cifar10_class_names = ['airplane', 'automobile', 'bird', 'cat', 'deer',
'dog', 'frog', 'horse', 'ship', 'truck']

    label_binarizer = LabelBinarizer()
    label_binarizer.fit(range(num_classes))
    label_inds = label_binarizer.inverse_transform(np.array(target_labels))

    fig, axies = plt.subplots(nrows=4, ncols=2)
    fig.tight_layout()
    fig.suptitle('Softmax Predictions', fontsize=20, y=1.1)

    num_predictions = 4
    margin = 0.05
    ind = np.arange(num_predictions)
    width = (1. - 2. * margin) / num_predictions

    for image_ind, (feature, label_ind, prediction_indicies, prediction_values) 
in enumerate(zip(input_features, label_inds, samples_predictions.indices, 
samples_predictions.values)):
        prediction_names = [cifar10_class_names[pred_i] for pred_i in 
prediction_indicies]
        correct_name = cifar10_class_names[label_ind]

        axies[image_ind][0].imshow(feature)
        axies[image_ind][0].set_title(correct_name)
        axies[image_ind][0].set_axis_off()

        axies[image_ind][1].barh(ind + margin, prediction_values[::-1], width)
        axies[image_ind][1].set_yticks(ind + margin)
        axies[image_ind][1].set_yticklabels(prediction_names[::-1])
        axies[image_ind][1].set_xticks([0, 0.5, 1.0])
```

现在,还原训练好的模型,并在测试集上测试它。

```python
test_batch_size = 64
save_model_path = './cifar-10_classification'
#Number of images to visualize
num_samples = 4

#Number of top predictions
top_n_predictions = 4

#Defining a helper function for testing the trained model
def test_classification_model():

    input_test_features, target_test_labels = pickle.load(open('preprocess_test.p', mode='rb'))
    loaded_graph = tf.Graph()
    with tf.Session(graph=loaded_graph) as sess:

        # loading the trained model
        model = tf.train.import_meta_graph(save_model_path + '.meta')
        model.restore(sess, save_model_path)

        # Getting some input and output Tensors from loaded model
        model_input_values = loaded_graph.get_tensor_by_name('input_images:0')
        model_target = loaded_graph.get_tensor_by_name('input_images_target:0')
        model_keep_prob = loaded_graph.get_tensor_by_name('keep_prob:0')
        model_logits = loaded_graph.get_tensor_by_name('logits:0')
        model_accuracy = loaded_graph.get_tensor_by_name('model_accuracy:0')

        # Testing the trained model on the test set batches
        test_batch_accuracy_total = 0
        test_batch_count = 0

        for input_test_feature_batch, input_test_label_batch in batch_split_features_labels(input_test_features, target_test_labels, test_batch_size):
            test_batch_accuracy_total += sess.run(
                model_accuracy,
                feed_dict={model_input_values: input_test_feature_batch, model_target: input_test_label_batch, model_keep_prob: 1.0})
            test_batch_count += 1

        print('Test set accuracy: {}\n'.format(test_batch_accuracy_total/test_batch_count))
```

第 8 章 目标检测——CIFAR-10 示例

```
# print some random images and their corresponding predictions from the
test set results
 random_input_test_features, random_test_target_labels =
tuple(zip(*random.sample(list(zip(input_test_features,
target_test_labels)), num_samples)))

random_test_predictions = sess.run(
  tf.nn.top_k(tf.nn.softmax(model_logits), top_n_predictions),
  feed_dict={model_input_values: random_input_test_features, model_target:
random_test_target_labels, model_keep_prob: 1.0})

display_samples_predictions(random_input_test_features,
random_test_target_labels, random_test_predictions)

#Calling the function
test_classification_model()
```

输出如下。

```
INFO:tensorflow:Restoring parameters from ./cifar-10_classification
Test set accuracy: 0.7540007961783439
```

输出结果如图 8.3 所示。

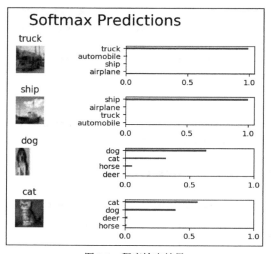

图 8.3　程序输出结果

接着再看可视化的另外一个例子，其中的误差如图 8.4 所示。

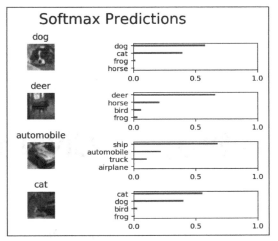

图 8.4 误差

现在，模型的测试准确率大约为 75%，这对于本节使用的这种简单的 CNN 来说不算太坏。

8.3 总结

本章展示了如何建立一个 CNN 来对 CIFAR-10 数据集中的图像进行分类。该模型在测试集上的分类准确率为 79%～80%。本章还绘制了卷积层的输出，但是从中很难看出神经网络是如何识别输入图像并对它们进行分类的，这可能需要用到更好的可视化技术。

接下来，本书将会介绍现代令人兴奋的深度学习实践之一——迁移学习。迁移学习允许程序在只有小型数据集时使用数据贪婪的深度学习架构。

第 9 章
目标检测——CNN 迁移学习

"个人如何从一个环境中转移到具有相似特征的另一个环境?"

——E. L. Thorndike, R. S. Woodworth (1991)

迁移学习(transfer learning,TL)是数据科学中研究的一个问题,主要涉及在解决特定任务时不断获取知识,并使用这些已获得的知识来解决另一个不同但相似的任务。本章将展示数据科学领域中使用 TL 的现代实践和共同主题之一。这里的想法是在处理具有较小数据集的领域中的问题时,如何从具有非常大的数据集的领域中来获得帮助。最后,本章将重新探讨 CIFAR-10 目标检测示例,并尝试使用 TL 来缩短训练时间和减小性能误差。

本章主要包括以下两个主题。

- 迁移学习。
- CIFAR-10 目标检测——回顾。

9.1 迁移学习

深度学习架构对于数据是贪婪的,在训练集中有一些样本不能够充分发挥作用。TL 通过将从解决大数据集任务中学习和获得的知识/表示迁移到另一个具有小数据集的不同但相似的任务上来解决这个问题。

TL 不但适用于小型数据集的情况,而且可以加速训练过程。从头开始训练大型的深度学习架构通常是非常慢的,因为可能需要学习数百万的权重参数。相反,我们可以考虑使用 TL 方法,将从类似问题中学习到的权重微调到自己尝试解决的问题上来。

9.1.1 迁移学习背后的直觉

这里通过使用教师-学生来类比 TL 背后的直觉。教师在他所熟悉的领域中有多年的教学经验，学生从教师讲授的课程中获得对该课程的一个简单理解。因此可以认为教师正在以简明扼要的方式向学生传授着知识。

同样，教师与学生的类比通常可以应用于通过深度学习或者神经网络来传递知识的情境中。模型学习到数据的一些特征，这些特征由网络的权重来表示。这些学习到的/特征（权重）能够转移到另一个不同但是相似的任务中。为了使深度学习架构收敛，将学习到的权重转移到另一个任务的过程将减少程序对大型数据集的需求，并且与从头开始训练模型相比，它还将缩短模型适应于新数据集所需的时间。

深度学习如今广泛应用，但通常大多数人在训练深度学习架构时使用 TL。很少有人从头开始训练深度学习架构，因为大多数时候很难有足够大的数据集来使深度学习模型收敛。因此，在诸如 ImageNet（该数据集拥有 120 万幅图像）这样的大型数据集上使用预训练的模型是很常见的，并将该训练模型应用于新的任务。可以使用预训练模型的权重参数作为特征提取器，或者可以使用它来初始化自己的模型架构，然后针对新任务对它们进行微调。使用 TL 的情况主要有 3 种，具体如下。

- **使用卷积网络作为固定的特征提取器**：在这种情况下，需使用在诸如 ImageNet 等大型数据集上预训练的卷积模型，然后让它与自己的问题相适应。例如，ImageNet 上的预训练卷积模型有一个全连接层，它输出的 ImageNet 中 1000 类物体的得分。所以需要删除该全连接层，因为这里不会再关心 ImageNet 的分类。然后，将所有其他层（即全连接层之前所有的层）视为一个特征提取器。一旦使用这个预训练模型提取出了特征，就可以将这些特征提供给任何的线性分类器，如 softmax 分类器，甚至线性 SVM 等。
- **微调卷积神经网络**：第二种情况也涉及第一种情况，但是需要额外的代价来使用反向传播算法在新任务上微调预训练权重参数。通常，我们都会保持大部分层固定不变，只微调网络的最后几层。调整整个网络或者网络的大部分层可能会导致过拟合。因此，我们可能只对调整那些与图像语义级特征相关联的层感兴趣。保持前面层固定不变背后的道理就是它们包含大多数图像任务中常见的通用或低级特征，如角落、边缘等。如果要引入预训练模型的原始数据集中不存在的新的类别，那么微调网络的高级层或顶端层是非常有用的，见图 9.1。

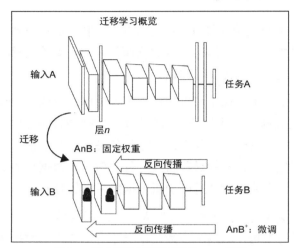

图 9.1　针对新任务微调预训练 CNN 模型

- **预训练模型**：第三种普遍的情况就是下载网上提供的模型检查点。如果没有很强的计算能力能够从头开始训练模型，那么可以选择这种方式，这样只需要使用别人发布的检查点来初始化模型，然后进行微调即可。

9.1.2　传统机器学习与迁移学习之间的不同

正如读者所见，应用机器学习的传统方式和涉及 TL 的机器学习之间存在明显的差异，如图 9.2 所示。传统的机器学习不会将任何知识或表示转移到任何其他任务中，但在 TL 中不是这样的。有时候，人们错误地运用了 TL，因此本节会给出一些条件，在这些条件下，使用 TL 能获得最大收益。

应用 TL 的条件如下。

- 与传统机器学习不同，源、目标任务或域不必来自相同的分布，但是它们必须相似。
- 如果训练样本较少或者没有必备的计算能力，也可以使用 TL。

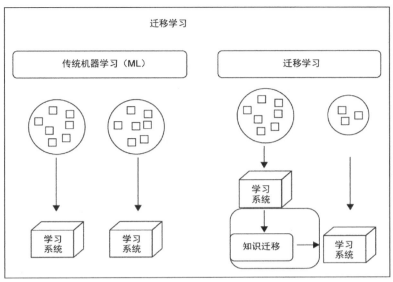

图 9.2 传统机器学习与使用 TL 的机器学习的对比

9.2 CIFAR-10 目标检测——回顾

在前面的内容中，我们在 CIFAR-10 数据集上训练了一个简单的**卷积神经网络（CNN）模型**。本节将会使用预训练模型作为特征提取器，同时移除预训练模型中的全连接层，然后将这些提取的特征或传递的值提供给 softmax 层。

在该实现中，预训练模型采用的是 inception 模型，它将在 ImageNet 上进行预训练。但是请记住，该实现是建立在引入 CNN 的前两节的基础上的。

9.2.1 解决方案大纲

同样地，这里将会替换预训练的 inception 模型中最后的全连接层，然后使用 inception 模型的其余部分作为特征提取器。因此，首先给 inception 模型提供原始图像，它将会从中提取出特征，然后输出所谓的迁移值。

在从 inception 模型中得到了已提取特征的迁移值之后，可能需要将它保存在桌面上，因为如果在运行中这样做是需要花费时间的，所以将它保存在桌面上可以节省时间。TensorFlow 教程中使用的是术语瓶颈值，而不是迁移值，但是这也只是同一个东西的不同名称而已。

在得到迁移值或者从桌面加载它们之后，可以将它们提供给任何为新任务定制的线性分类器。这里，我们会把提取到的迁移值提供给另外一个神经网络，然后训练 CIFAR-10 数据集中新的类别。

图 9.3 所示为后续将会遵循的一般解决方案大纲。

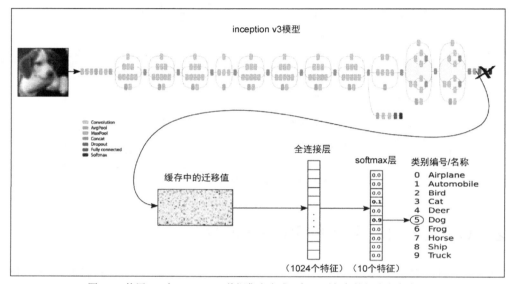

图 9.3　使用 TL 在 CIFAR-10 数据集上完成目标识别任务的解决方案大纲

9.2.2　加载和探索 CIFAR-10 数据集

本节从导入此实现所需的软件包开始。

```
%matplotlib inline
import matplotlib.pyplot as plt
import tensorflow as tf
import numpy as np
import time
from datetime import timedelta
import os

# Importing a helper module for the functions of the Inception model.
import inception
```

接下来，加载另一个辅助脚本，它可以用来下载处理过的 CIFAR-10 数据集。

```
import cifar10
#importing number of classes of CIFAR-10
from cifar10 import num_classes
```

如果还没有这样做过,那么需要设置 CIFAR-10 数据集的路径。cifar-10.py 脚本将使用该路径来保存数据集。

```
cifar10.data_path = "data/CIFAR-10/"
```

```
The CIFAR-10 dataset is about 170 MB, the next line checks if the dataset
is already downloaded if not it downloads the dataset and store in the
previous data_path:
```

```
cifar10.maybe_download_and_extract</span>()
```

输出如下。

```
- Download progress: 100.0%
Download finished. Extracting files.
Done.
```

下面查看 CIFAR-10 数据集中有哪些类别。

```
#Loading the class names of CIFAR-10 dataset
class_names = cifar10.load_class_names()
class_names
```

输出如下。

```
Loading data: data/CIFAR-10/cifar-10-batches-py/batches.meta
['airplane',
 'automobile',
 'bird',
 'cat',
 'deer',
 'dog',
 'frog',
 'horse',
 'ship',
 'truck']
Load the training-set.
```

以上代码会返回图像、整型的类别数字和称为标签的独热编码数组表示的类别数字。

```
training_images, training_cls_integers, trainig_one_hot_labels =
cifar10.load_training_data()
```

输出如下。

```
Loading data: data/CIFAR-10/cifar-10-batches-py/data_batch_1
Loading data: data/CIFAR-10/cifar-10-batches-py/data_batch_2
Loading data: data/CIFAR-10/cifar-10-batches-py/data_batch_3
Loading data: data/CIFAR-10/cifar-10-batches-py/data_batch_4
Loading data: data/CIFAR-10/cifar-10-batches-py/data_batch_5
Load the test-set.
```

下面对测试集做同样的操作，方法是加载图像数据和它们对应的带有独热编码的目标类别的整数表示方式。

```
#Loading the test images, their class integer, and their corresponding one-hot
encoding
testing_images, testing_cls_integers, testing_one_hot_labels =
cifar10.load_test_data()
```

输出如下。

```
Loading data: data/CIFAR-10/cifar-10-batches-py/test_batch
```

然后，查看 CIFAR-10 数据集中训练集和测试集的分布。

```
print("-Number of images in the training
set:\t\t{}".format(len(training_images)))
print("-Number of images in the testing
set:\t\t{}".format(len(testing_images)))
```

输出如下。

```
-Number of images in the training set:          50000
-Number of images in the testing set:           10000
```

接下来，定义一些辅助函数用于研究数据集。下面的辅助函数在网格中绘制了 9 幅图像。

```
def plot_imgs(imgs, true_class, predicted_class=None):

    assert len(imgs) == len(true_class)
```

9.2 CIFAR-10目标检测——回顾

```
# Creating a placeholders for 9 subplots
fig, axes = plt.subplots(3, 3)
# Adjustting spacing.
if predicted_class is None:
    hspace = 0.3
else:
    hspace = 0.6
fig.subplots_adjust(hspace=hspace, wspace=0.3)

for i, ax in enumerate(axes.flat):
    # There may be less than 9 images, ensure it doesn't crash.
    if i < len(imgs):
        # Plot image.
        ax.imshow(imgs[i],
                  interpolation='nearest')

        # Get the actual name of the true class from the class_names array
        true_class_name = class_names[true_class[i]]

        # Showing labels for the predicted and true classes
        if predicted_class is None:
            xlabel = "True: {0}".format(true_class_name)
        else:
            # Name of the predicted class.
            predicted_class_name = class_names[predicted_class[i]]

            xlabel = "True: {0}\nPred: {1}".format(true_class_name,
predicted_class_name)

        ax.set_xlabel(xlabel)
    # Remove ticks from the plot.
    ax.set_xticks([])
    ax.set_yticks([])
plt.show()
```

接下来，继续观察并可视化测试集中的一些图像及其对应的实际类别。

```
# get the first 9 images in the test set
imgs = testing_images[0:9]
```

```
# Get the integer representation of the true class.
true_class = testing_cls_integers[0:9]

# Plotting the images
plot_imgs(imgs=imgs, true_class=true_class)
```

输出如图 9.4 所示。

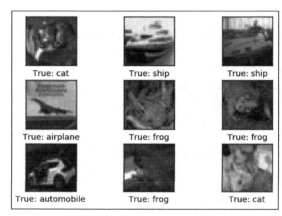

图 9.4　测试集中的前 9 幅图像

9.2.3　inception 模型迁移值

正如前面提到的，这里会用到在 ImageNet 数据集上预训练的 inception 模型。所以，需要从网上下载预训练模型。

现在，开始为 inception 模型定义 data_dir。

```
inception.data_dir = 'inception/'
```

预训练的 inception 模型的权重参数大概为 85MB。如果在前面定义的 data_dir 路径下找不到该权重，那么如下代码将会重新下载它。

```
inception.maybe_download()

Downloading Inception v3 Model ...
- Download progress: 100%
```

接下来，加载 inception 模型，然后将它作为 CIFAR-10 图像的特征提取器。

```
# Loading the inception model so that we can inialized it with the pre-trained
weights and customize for our model
inception_model = inception.Inception()
```

如前所述，计算 CIFAR-10 数据集的迁移值将会花费一定时间，所以需要将它们缓存下来以备后续使用。幸运的是，inception 模块中有一个辅助函数可以帮助我们做到这一点。

```
from inception import transfer_values_cache
```

接下来，需要为缓存的训练和测试文件设置文件路径。

```
file_path_train = os.path.join(cifar10.data_path,
'inception_cifar10_train.pkl')
file_path_test = os.path.join(cifar10.data_path,
'inception_cifar10_test.pkl')
print("Processing Inception transfer-values for the training images of
Cifar-10 ...")
# First we need to scale the imgs to fit the Inception model requirements
as it requires all pixels to be from 0 to 255,
# while our training examples of the CIFAR-10 pixels are between 0.0 and
1.0
imgs_scaled = training_images * 255.0

# Checking if the transfer-values for our training images are already
calculated and loading them, if not calculate and save them.
transfer_values_training =
transfer_values_cache(cache_path=file_path_train,
                                        images=imgs_scaled,
                                        model=inception_model)
print("Processing Inception transfer-values for the testing images of
Cifar-10 ...")
# First we need to scale the imgs to fit the Inception model requirements
as it requires all pixels to be from 0 to 255,
# while our training examples of the CIFAR-10 pixels are between 0.0 and
1.0
imgs_scaled = testing_images * 255.0
# Checking if the transfer-values for our training images are already
calculated and loading them, if not calcaulate and save them.
transfer_values_testing = transfer_values_cache(cache_path=file_path_test,
                                  images=imgs_scaled,
                                  model=inception_model)
```

前面提到了，在 CIFAR-10 数据集的训练集中有 50000 张图像。下面查看这些图像迁移值的形状。此训练集中每幅图像的迁移值形状都应该是 2048。

```
transfer_values_training.shape
```

输出如下。

```
(50000, 2048)
```

这里需要对测试集做同样的操作。

```
transfer_values_testing.shape
```

输出如下。

```
(10000, 2048)
```

为了直观地理解迁移值的形状,这里将会定义一个辅助函数,该函数可以绘制训练集或者测试集中某幅特定图像的迁移值。

```
def plot_transferValues(ind):
    print("Original input image:")
    # Plot the image at index ind of the test set.
    plt.imshow(testing_images[ind], interpolation='nearest')
    plt.show()

    print("Transfer values using Inception model:")
    # Visualize the transfer values as an image.
    transferValues_img = transfer_values_testing[ind]
    transferValues_img = transferValues_img.reshape((32, 64))

    # Plotting the transfer values image.
    plt.imshow(transferValues_img, interpolation='nearest', cmap='Reds')
    plt.show()
plot_transferValues(i=16)
```

输入图像如图 9.5 所示。

使用 inception 模型计算该图像的迁移值,如图 9.6 所示。

```
plot_transferValues(i=17)
```

输入图像如图 9.7 所示。

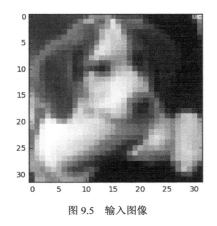

图 9.5　输入图像

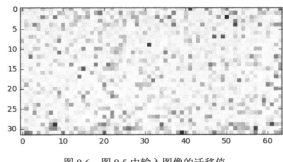

图 9.6　图 9.5 中输入图像的迁移值

使用 inception 模型计算该图像的迁移值，结果如图 9.8 所示。

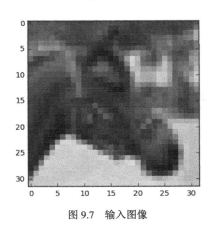

图 9.7　输入图像

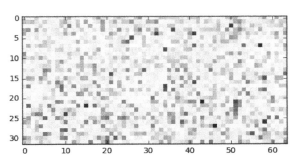

图 9.8　图 9.7 中输入图像的迁移值

9.2.4　迁移值分析

本节将会对之前从训练图像中获得的迁移值进行一些分析。做这些分析的目的是判断这些迁移值是否已经足够用于对 CIFAR-10 数据集中的图像进行分类。

对于每一幅输入图像获取了维度是 2048 的迁移值。为了绘制出这些迁移值并对它们做进一步分析，可以使用降维法，如 scikit-learn 包中的**主成分分析**（Principal Component Analysis，PCA）方法。这里将迁移值从 2048 维降低到 2 维以便对它们进行可视化，并查看它们是否是区分 CIFAR-10 数据集中不同类别的好的特征。

```
from sklearn.decomposition import PCA
```

接下来，需要创建一个 PCA 对象，其中的主成分数目为 2。

```
pca_obj = PCA(n_components=2)
```

因为将迁移值从 2048 维降低到 2 维需要花费大量时间，所以这里将从具有迁移值的 5000 幅图像中抽出 3000 幅图像，并作为一个子集。

```
subset_transferValues = transfer_values_training[0:3000]
```
还需要获取这些图像的类别数字。

```
cls_integers = testing_cls_integers[0:3000]
```

通过输出迁移值的形状，可以再次检查抽出的子集是否正确。

```
subset_transferValues.shape
```

输出如下。

```
(3000, 2048)
```

接下来，使用前面定义的 PCA 对象来将迁移值从 2048 维降低到 2 维。

```
reduced_transferValues = pca_obj.fit_transform(subset_transferValues)
```

现在，查看 PCA 降维过程的输出。

```
reduced_transferValues.shape
```

输出如下。

```
(3000, 2)
```

在将迁移值的维度降低到只有 2 之后，再绘制出这些值。

```
#Importing the color map for plotting each class with different color.
import matplotlib.cm as color_map

def plot_reduced_transferValues(transferValues, cls_integers):
    # Create a color-map with a different color for each class.
    c_map = color_map.rainbow(np.linspace(0.0, 1.0, num_classes))
    
    # Getting the color for each sample.
```

```
colors = c_map[cls_integers]

# Getting the x and y values.
x_val = transferValues[:, 0]
y_val = transferValues[:, 1]

# Plot the transfer values in a scatter plot
plt.scatter(x_val, y_val, color=colors)
plt.show()
```

下面我们将绘制来自训练集的子集降维之后的迁移值。在 CIFAR-10 数据集中有 10 个类别，因此将使用不同的颜色来绘制相应的迁移值。从图 9.9 中可以看出，迁移值根据相应的类别进行了分组。不同组之间存在重叠是因为 PCA 降维过程无法完全正确地分离迁移值。

```
plot_reduced_transferValues(reduced_transferValues, cls_integers)
```

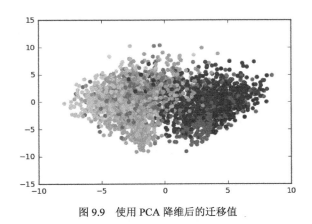

图 9.9　使用 PCA 降维后的迁移值

还可以使用名为 t-SNE 的不同降维法来对迁移值进行进一步分析。

```
from sklearn.manifold import TSNE
```

同样地，还需要对 2048 维的迁移值进行降维，但是这里降到 50 维，而不是 2 维。

```
pca_obj = PCA(n_components=50)
transferValues_50d = pca_obj.fit_transform(subset_transferValues)
```

接下来，将会叠加第二个降维法，以 PCA 处理后的输出作为输入。

```
tsne_obj = TSNE(n_components=2)
```

最后，将 t-SNE 方法应用于使用 PCA 方法获得的降维后的值。

```
reduced_transferValues = tsne_obj.fit_transform(transferValues_50d)
```

同时再次检查降维后的值是否具有正确的形状。

```
reduced_transferValues.shape
```

输出如下。

```
(3000, 2)
```

现在绘制 t-SNE 方法降维之后的迁移值。如图 9.10 所示，t-SNE 方法能够比 PCA 方法更好地分离不同组的迁移值。

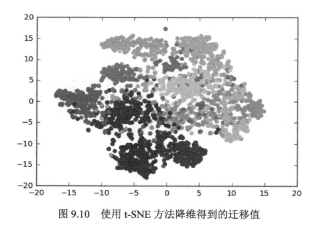

图 9.10　使用 t-SNE 方法降维得到的迁移值

从这个分析中得出的结论是：通过将输入图像输入预训练的 inception 模型中，提取出来的迁移值能够用来将训练图像分成 10 个类。这个分离不会 100%正确，因为不同类别之间存在小的重叠，但是可以通过对预训练模型进行一些微调来消除这种重叠。

```
plot_reduced_transferValues(reduced_transferValues, cls_integers)
```

现在有了从训练图像中提取出来的迁移值，并且知道这些值能够在某种程度上区分 CIFAR-10 数据集中不同的类别。接下来，需要构建一个线性分类器，并将这些迁移值输入给线性分类器，以便进行实际的分类。

9.2.5 模型构建与训练

首先,指定将会被输入神经网络模型中的输入占位符变量。第一个输入变量(包含提取的迁移值)的形状为[None, transfer_len]。第二个占位符变量将以独热向量的格式保存训练集的实际类别标签。

```
transferValues_arrLength = inception_model.transfer_len
input_values = tf.placeholder(tf.float32, shape=[None,
transferValues_arrLength], name='input_values')
y_actual = tf.placeholder(tf.float32, shape=[None, num_classes],
name='y_actual')
```

还可以定义另一个占位符变量来获取每一个类别的相应整数值(1~10)。

```
y_actual_cls = tf.argmax(y_actual, axis=1)
```

接下来,需要构建实际的分类神经网络,它将接受这些输入占位符并生成预测类别。

```
def new_weights(shape):
    return tf.Variable(tf.truncated_normal(shape, stddev=0.05))

def new_biases(length):
    return tf.Variable(tf.constant(0.05, shape=[length]))

def new_fc_layer(input,          # The previous layer.
                num_inputs,      # Num. inputs from prev. layer.
                num_outputs,     # Num. outputs.
                use_relu=True):  # Use Rectified Linear Unit (ReLU)?

    # Create new weights and biases.
    weights = new_weights(shape=[num_inputs, num_outputs])
    biases = new_biases(length=num_outputs)

    # Calculate the layer as the matrix multiplication of
    # the input and weights, and then add the bias-values.
    layer = tf.matmul(input, weights) + biases

    # Use ReLU?
    if use_relu:
        layer = tf.nn.relu(layer)

    return layer
```

```
# First fully-connected layer.
layer_fc1 = new_fc_layer(input=input_values,
                         num_inputs=2048,
                         num_outputs=1024,
                         use_relu=True)

# Second fully-connected layer.
layer_fc2 = new_fc_layer(input=layer_fc1,
                         num_inputs=1024,
                         num_outputs=num_classes,
                         use_relu=False)

# Predicted class-label.
y_predicted = tf.nn.softmax(layer_fc2)

# Cross-entropy for the classification of each image.
cross_entropy = \
    tf.nn.softmax_cross_entropy_with_logits(logits=layer_fc2,
                                            labels=y_actual)

# Loss aka. cost-measure.
# This is the scalar value that must be minimized.
loss = tf.reduce_mean(cross_entropy)
```

然后，需要定义分类器训练期间使用的优化标准。在此实现中，我们将会使用 **AdamOptimizer**。分类器的输出将是含有 10 个概率分数的数组，这些数组对应 CIFAR-10 数据集中的类别数量。接下来，将会在此数组上执行 **argmax** 操作，来给出输入样本具有最大概率值的类别。

```
step = tf.Variable(initial_value=0,
                   name='step', trainable=False)
optimizer = tf.train.AdamOptimizer(learning_rate=1e-4).minimize(loss, step)
y_predicted_cls = tf.argmax(y_predicted, axis=1)
#compare the predicted and true classes
correct_prediction = tf.equal(y_predicted_cls, y_actual_cls)
#cast the boolean values to fload
model_accuracy = tf.reduce_mean(tf.cast(correct_prediction, tf.float32))
```

接下来，需要定义一个实际执行图计算的 TensorFlow 会话，然后初始化此实现中前面定义的变量。

```python
session = tf.Session()
session.run(tf.global_variables_initializer())
```

在此实现中，我们将会使用**随机梯度下降**（SGD）法，所以需要定义一个函数，以从含有 50000 个图像的训练集中随机生成特定大小的批量数据。

因此，还需要定义一个辅助函数用于从迁移值的输入训练集中生成一批随机数据。

```python
#defining the size of the train batch
train_batch_size = 64

#defining a function for randomly selecting a batch of images from the
dataset
def select_random_batch():
    # Number of images (transfer-values) in the training-set.
    num_imgs = len(transfer_values_training)

    # Create a random index.
    ind = np.random.choice(num_imgs,
                           size=training_batch_size,
                           replace=False)

    # Use the random index to select random x and y-values.
    # We use the transfer-values instead of images as x-values.
    x_batch = transfer_values_training[ind]
    y_batch = trainig_one_hot_labels[ind]

    return x_batch, y_batch
```

接下来，需要定义一个辅助函数来执行实际的优化过程，它将用于优化网络的权重。它将在每一轮迭代时生成一个批数据，并基于该批数据来优化网络。

```python
def optimize(num_iterations):

    for i in range(num_iterations):
        # Selectin a random batch of images for training
        # where the transfer values of the images will be stored in
input_batch
        # and the actual labels of those batch of images will be stored in
y_actual_batch
        input_batch, y_actual_batch = select_random_batch()
```

```
        # storing the batch in a dict with the proper names
        # such as the input placeholder variables that we define above.
        feed_dict = {input_values: input_batch,
                      y_actual: y_actual_batch}

        # Now we call the optimizer of this batch of images
        # TensorFlow will automatically feed the values of the dict we
created above
        # to the model input placeholder variables that we defined above.
        i_global, _ = session.run([step, optimizer],
                                   feed_dict=feed_dict)

        # print the accuracy every 100 steps.
        if (i_global % 100 == 0) or (i == num_iterations - 1):
            # Calculate the accuracy on the training-batch.
            batch_accuracy = session.run(model_accuracy,
                                          feed_dict=feed_dict)

            msg = "Step: {0:>6}, Training Accuracy: {1:>6.1%}"
            print(msg.format(i_global, batch_accuracy))
```

然后，定义一些辅助函数来显示之前神经网络的结果，同时显示预测结果的混淆矩阵。

```
def plot_errors(cls_predicted, cls_correct):
    # cls_predicted is an array of the predicted class-number for
    # all images in the test-set.

    # cls_correct is an array with boolean values to indicate
    # whether is the model predicted the correct class or not.

    # Negate the boolean array.
    incorrect = (cls_correct == False)
    # Get the images from the test-set that have been
    # incorrectly classified.
    incorrectly_classified_images = testing_images[incorrect]
    # Get the predicted classes for those images.
    cls_predicted = cls_predicted[incorrect]

    # Get the true classes for those images.
    true_class = testing_cls_integers[incorrect]
```

```
    n = min(9, len(incorrectly_classified_images))
    # Plot the first n images.
    plot_imgs(imgs=incorrectly_classified_images[0:n],
              true_class=true_class[0:n],
              predicted_class=cls_predicted[0:n])
```

接下来,需要定义辅助函数来绘制混淆矩阵。

```
from sklearn.metrics import confusion_matrix

def plot_confusionMatrix(cls_predicted):

    # cls_predicted array of all the predicted
    # classes numbers in the test.

    # Call the confucion matrix of sklearn
    cm = confusion_matrix(y_true=testing_cls_integers,
                          y_pred=cls_predicted)

    # Printing the confusion matrix
    for i in range(num_classes):
        # Append the class-name to each line.
        class_name = "({}) {}".format(i, class_names[i])
        print(cm[i, :], class_name)

    # labeling each column of the confusion matrix with the class number
    cls_numbers = [" ({0})".format(i) for i in range(num_classes)]
    print("".join(cls_numbers))
```

此外,还要定义另一个辅助函数来在测试集上运行训练的分类器,并衡量训练模型在测试集上的准确率。

```
# Split the data-set in batches of this size to limit RAM usage.
batch_size = 128

def predict_class(transferValues, labels, cls_true):
    # Number of images.
    num_imgs = len(transferValues)

    # Allocate an array for the predicted classes which
```

```
    # will be calculated in batches and filled into this array.
    cls_predicted = np.zeros(shape=num_imgs, dtype=np.int)

    # Now calculate the predicted classes for the batches.
    # We will just iterate through all the batches.
    # There might be a more clever and Pythonic way of doing this.

    # The starting index for the next batch is denoted i.
    i = 0

    while i < num_imgs:
        # The ending index for the next batch is denoted j.
        j = min(i + batch_size, num_imgs)

        # Create a feed-dict with the images and labels
        # between index i and j.
        feed_dict = {input_values: transferValues[i:j],
                     y_actual: labels[i:j]}

        # Calculate the predicted class using TensorFlow.
        cls_predicted[i:j] = session.run(y_predicted_cls,
feed_dict=feed_dict)

        # Set the start-index for the next batch to the
        # end-index of the current batch.
        i = j
    # Create a boolean array whether each image is correctly classified.
    correct = [a == p for a, p in zip(cls_true, cls_predicted)]

    return correct, cls_predicted

#Calling the above function making the predictions for the test

def predict_cls_test():
    return predict_class(transferValues = transfer_values_test,
                         labels = labels_test,
                         cls_true = cls_test)

def classification_accuracy(correct):
    # When averaging a boolean array, False means 0 and True means 1.
    # So we are calculating: number of True / len(correct) which is
```

```python
    # the same as the classification accuracy.

    # Return the classification accuracy
    # and the number of correct classifications.
    return np.mean(correct), np.sum(correct)

def test_accuracy(show_example_errors=False,
                  show_confusion_matrix=False):

    # For all the images in the test-set,
    # calculate the predicted classes and whether they are correct.
    correct, cls_pred = predict_class_test()
    # Classification accuracypredict_class_test and the number of correct classifications.
    accuracy, num_correct = classification_accuracy(correct)
    # Number of images being classified.
    num_images = len(correct)

    # Print the accuracy.
    msg = "Test set accuracy: {0:.1%} ({1} / {2})"
    print(msg.format(accuracy, num_correct, num_images))

    # Plot some examples of mis-classifications, if desired.
    if show_example_errors:
        print("Example errors:")
        plot_errors(cls_predicted=cls_pred, cls_correct=correct)

    # Plot the confusion matrix, if desired.
    if show_confusion_matrix:
        print("Confusion Matrix:")
        plot_confusionMatrix(cls_predicted=cls_pred)
```

在进行任何优化之前,先看一下前面的神经网络模型的性能。

```
test_accuracy(show_example_errors=True,
              show_confusion_matrix=True)

Accuracy on Test-Set: 9.4% (939 / 10000)
```

可以看到,神经网络模型的性能很差,但是在根据我们之前定义的优化标准进行一些优化之后,它将会变得好一些。接下来将会运行优化器进行 10000 次迭代,并在此之后再次测试模型准确率。

```
optimize(num_iterations=10000)
test_accuracy(show_example_errors=True,
              show_confusion_matrix=True)
Accuracy on Test-Set: 90.7% (9069 / 10000)
Example errors:
```

结果输出如图 9.11 和以下代码所示。

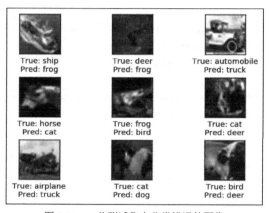

图 9.11　一些测试集中分类错误的图像

```
Confusion Matrix:
[926    6   13    2    3    0    1    1   29   19] (0) airplane
[  9  921    2    5    0    1    1    1    2   58] (1) automobile
[ 18    1  883   31   32    4   22    5    1    3] (2) bird
[  7    2   19  855   23   57   24    9    2    2] (3) cat
[  5    0   21   25  896    4   24   22    2    1] (4) deer
[  2    0   12   97   18  843   10   15    1    2] (5) dog
[  2    1   16   17   17    4  940    1    2    0] (6) frog
[  8    0   10   19   28   14    1  914    2    4] (7) horse
[ 42    6    1    4    1    0    2    0  932   12] (8) ship
[  6   19    2    2    1    0    1    1    9  959] (9) truck
 (0)  (1)  (2)  (3)  (4)  (5)  (6)  (7)  (8)  (9)
```

最后，关闭已打开的会话。

```
model.close()
session.close()
```

9.3 总结

本章介绍了最广泛使用的深度学习最佳实践之一。迁移学习（TL）是一个令人兴奋的工具，我们可以使用它来让深度学习架构从小数据集中学习，但是请确保以正确的方式来使用。

接下来，我们将会介绍一种在自然语言处理中广泛使用的深度学习架构。这种循环类型的架构在大多数 NLP 领域取得了突破，如机器翻译、语音识别、语言建模和情感分析等。

第 10 章
循环神经网络——语言模型

循环神经网络（recurrent neural network，RNN）是一种深度学习架构，它广泛地应用在自然语言处理（natural language processing，NLP）中。这类架构使得开发人员能够为当前的预测提供上下文信息，并且一些特别的架构可以在任何输入序列中处理长期依赖问题。本章将展示如何构建一个序列到序列的模型，它将在许多 NLP 应用中发挥作用。本章将通过构建一个字符级的语言模型来解释这些概念，并让读者了解模型如何产生和原输入序列相似的句子。

本章将包含以下主题。
- RNN 的直观解释。
- LSTM 网络。
- 语言模型的实现。

10.1 RNN 的直观解释

到目前为止，本书介绍的所有深度学习架构都无法记住它们之前接收的输入信息。比如，如果给一个**前馈神经网络（FNN）**输入如 HELLO 这样的一串字符，当网络接收到 E 的时候，它并没有保留（或者说遗忘了）它刚刚读取了 H 的信息。这对于基于序列的学习来说是一个很严重的问题。同时由于它没有任何它之前读取的字母的记忆，因此这类网络很难用来预测下一个字符。这对诸如语言建模、机器翻译、语音识别等应用是没有意义的。

出于这个原因，我们引入了 RNN，RNN 是一组能够保留并记住之前遇到过的信息的深度学习架构。

现在解释一下 RNN 是怎么在相同的字符序列 **HELLO** 上工作的。当 RNN 细胞（单元）接收到 **E** 这个输入时，它也接收到了它早先接收到的字符 **H**。将现在的字符与过去的字符一起作为 RNN 细胞的输入给这些架构带来了很大的好处，即短期记忆，还使得这些架构可用于在这个特定的字符序列中预测（猜测）**H** 之后最可能出现的字符，即 **L**。

之前提到的架构会为输入分配权重；RNN 遵循相同的优化过程为其多个输入（当前输入和过去输入）分配权重。因此，在这种情况下，网络将为它们中的每一个输入分配一个权重矩阵。为了做到这一点，我们将使用**梯度下降法**和较复杂版本的反向传播网络，该反向传播网络又叫作**时间反向传播**（backpropagation through time，BPTT）网络。

10.1.1　RNN 的架构

根据之前使用的深度学习架构的背景，读者将了解到 RNN 的特殊之处。之前所学习的架构在输入和训练方面并不灵活。它们以大小固定的序列、向量或图像作为输入，并以另一个大小固定的序列、向量或图像作为输出。RNN 架构略有不同，它使得开发人员能够将序列作为输入并输出另一个序列，或者仅仅在输入或输出时使用序列，如图 10.1 所示。这种灵活性对诸如语言模型和情感分析这类的应用来说非常有用。

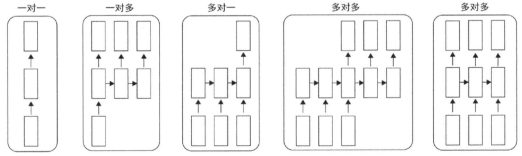

图 10.1　RNN 在输入或输出形状方面的灵活性（在 GitHub 网站中搜索 "The Unreasonable Effectiveness of Recurrent Neural Networks" 即可查看该图片）

这类架构背后的直观解释是模仿人类处理信息的方式。在任何典型的谈话中，对某人的话语的理解完全取决于他之前所说的内容，甚至可以根据他刚才所说的内容预测他接下来会说些什么。

在 RNN 的情况下，我们应该遵循完全相同的过程。例如，要翻译句子当中的某个特定词语，我们不能用传统的 FNN，因为它们无法将之前的单词翻译和当前想要翻译的单词一起作为输入，这可能会造成由于缺少单词的上下文信息而错误翻译的情况。

RNN 确实保留了有关过去的信息，并且它们会用某种循环来使得过去学习的信息能够在任意给定点用于当前的预测。

在图 10.2 中，有名为 A 的神经网络，它接收输入 X_t 并产生输出 h_t。此外，它在循环结构的帮助下接收之前步骤产生的信息。

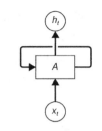

图 10.2　RNN 架构具有循环结构，能保留过去步骤的信息（在 GitHub 网站中搜索"Understanding LSTM Networks"即可查看该图片）

这个循环看起来似乎不清晰，但如果用图 10.2 所示的展开版本，就会发现它是十分简单而直观的，而 RNN 只是同一网络（可以是普通的 FNN）的重复版本，如图 10.3 所示。

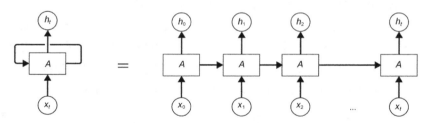

图 10.3　循环神经网络架构的展开版本（在 GitHub 网站中搜索"Understanding LSTM Networks"即可查看该图片）

这种直观的 RNN 架构及其在输入、输出形状方面的灵活性使得它非常适合基于序列的有趣学习任务，如机器翻译、语言建模、情感分析、图像标注等。

10.1.2　RNN 的示例

现在，我们已经对 RNN 如何工作以及它如何应用于不同的基于序列的有趣例子有了一个直观的了解。下面继续仔细了解其中一些有趣的例子。

1．字符级语言模型

语言建模对诸如语音识别、机器翻译等更多应用来说是一个重要的任务。本节将尝试模仿 RNN 的训练过程，并深入了解这些网络的工作原理。我们将构建一个对字符进行操作的语言模型。因此，我们将为模型提供很多文本，目的是在给出前一个字符的前提下构造出下一个字符的概率分布，这将允许开发人员生成类似于训练过程的那些输入的文本。

例如，假设有一种语言，它只以 4 个字母作为它的词汇——**helo**。任务是在一个特定的字符序列（如 hello）上训练一个循环神经网络。在这个特例中，有如下 4 个训练样本。

- 当给定第一个输入字符 **h** 的上下文时计算字符 **e** 的概率。
- 当给定 **he** 的上下文时计算字符 **l** 的概率。
- 当给定 **hel** 的上下文时计算字符 **l** 的概率。
- 当给定 **hell** 的上下文时计算字符 **o** 的概率。

正如之前所讲到的，作为机器学习的一部分，深度学习通常只以实数数值作为输入。所以，我们需要用某种方式将输入字符转换或编码成数值的形式。为了做到这一点，我们可以使用独热向量编码。独热向量是一种编码文本的方式，该向量中除了一个位置外，其余位置全为 0，不为 0 的这个分量就是所要建模的语言词汇（本例中就是 **helo**）中字符的索引。在编码完训练样本后，将它们依次提供给 RNN 类型的模型。在每个给定字符处，RNN 类型模型的输出将是 4 维向量（向量的大小对应于词汇的大小），它表示词汇当中的每个字符作为下一个字符出现在给定输入字符后面的概率。图 10.4 阐明了这一过程。

如图 10.4 所示，将输入序列的第一个字符 h 输入模型中，输出是一个 4 维向量，表示下一个字符的置信度。因此，h 以 1.0 的置信度作为输入 h 的下一个字符，e 以 2.2 的置信度作为下一个字符，l 以-3.0 的置信度作为下一个字符，而 o 以 4.1 的置信度作为下一个字符。在这个特定的例子中，正确的下一个字符是 e，这是根据训练序列 hello 得出的。所以在训练这个 RNN 时的主要目的是增大 e 作为下一个字符的置信度，同时减小其他字符的置信度。为了实现这种优化，我们将使用梯度下降算法和反向传播算法来更新权重并影响网络，以使它在正确的下一个字符 e 以及另外 3 个训练样本

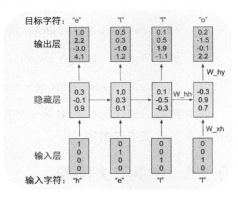

图 10.4 使用独热向量编码字符作为输入的 RNN 的示例，输出是词汇上的分布，表示在当前字符后最可能的字符（在 GitHub 网站中搜索"The Unreasonable Effectiveness of Recurrent Neural Networks"即可查看该图片）

上产生一个较高的置信度。

可以看到，RNN 的输出产生了一个词汇里所有字符成为下一个字符的置信度分布。置信度分布可以转换为概率分布，这样增大一个字符成为下一个字符的概率，就会导致减小其他字符成为下一个字符的概率，因为概率和必须为 1。对于这种特定的修改，可以对每个输出向量使用标准的 softmax 层。

为了从这类网络中产生文本，可以向模型中输入一个起始字符，并获得哪些字符将成为下一个字符的概率分布，然后从这些字符中进行采样，并将采样值作为输入输反馈到模型当中。通过一遍遍地重复这一过程，我们就可以获得一系列字符，重复次数和产生具有期望长度的文本所要求的次数一样。

2．用莎士比亚数据集的语言模型

在前面的例子中，我们可以获得一个产生文本的模型。但令人吃惊的是，网络不仅能够产生文本，它还能够学习训练数据中的风格和结构。这一有趣的过程可以通过在一类特定文本上训练一个 RNN 模型来表述，该类文本有自己的结构和风格，比如，下面莎士比亚的作品。

下面是训练好的网络产生的输出。

Second Senator:

They are away this miseries, produced upon my soul,

Breaking and strongly should be buried, when I perish

The earth and thoughts of many states.

尽管网络只知道每次产生单个字符，但它能够产生一串具有莎士比亚作品结构和风格的有意义的文本与名字。

10.1.3　梯度消失问题

当训练这类 RNN 架构时，我们要用到梯度下降算法和时间反向传播算法，这两种算法在许多基于序列的学习任务中取得了成功。但是由于梯度的性质和使用快速训练的策略，可以看到梯度往往变得太小甚至消失。这个过程引入了许多从业者会陷入的梯度消失问题，如图 10.5 所示。本章后面将讨论研究人员如何解决这类问题，并提出一般 RNN 的变种来解决这个问题。

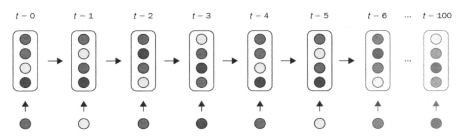

图 10.5 梯度消失问题

10.1.4 长期依赖问题

研究者面临的另一个挑战是长期依赖问题，人们可以在文本中遇到该类问题。例如，有人提供一个序列"I used to live in France and I learned how to speak…"，很明显序列的下一个词是"French"。

一般的 RNN 可以处理该种情况，因为 RNN 有短期依赖能力，如图 10.6 所示。

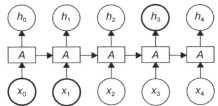

在另一个例子中，如果有人这样开始一个序列"I used to live in France…"然后他开始描述在那里美好的生活，最后他这样结束序列"I learned to speak French"。为了使模型在序列的结尾预测他所学的语言，模型必须保留较早的单词"live"和

图 10.6 文本中的短期依赖关系（在 GitHub 网站中搜索"Understanding LSTM Networks"即可查看该图片）

"France"的一些信息。如果模型无法跟踪文本中的长期依赖关系，则该模型将无法处理这种情况，如图 10.7 所示。

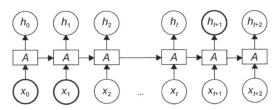

图 10.7 文本中的长期依赖问题（在 GitHub 网站中搜索
"Understanding LSTM Networks"即可查看该图片）

为了处理文本中的梯度消失和长期依赖问题，研究者提出了一种一般 RNN 的变种，称作**长短期记忆（Long Short Term Memory，LSTM）网络**。

10.2 LSTM 网络

作为 RNN 的一个变种,LSTM 常常用来解决文本中的长期依赖问题。LSTM 最初由 Hochreiter 和 Schmidhuber（1997）提出（见论文"LONG SHORT-TERM MEMORY"），很多研究者致力于研究它，并在许多领域产生了很多有趣的结果。

这类架构因为其内部架构能够处理文本中的长期依赖问题。

LSTM 网络和普通 RNN 类似，前者有一个在时间上重复的模块，但这个重复模块的内部架构和普通 RNN 有所不同。它包含了更多的层来遗忘和更新信息，如图 10.8 所示。

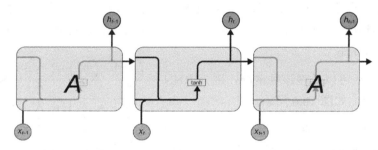

图 10.8　标准 RNN 中的重复模块包含单个层（在 GitHub 网站中搜索"Understanding LSTM Networks"即可查看该图片）

正如前面提到的，普通 RNN 有单个 NN 层，但是 LSTM 网络有 4 个不同的层并以一种特别的方式交互，如图 10.9 所示。在构建语言模型示例时我们将了解到，这种特殊的交互方式使得 LSTM 在很多领域都能很好地发挥作用。

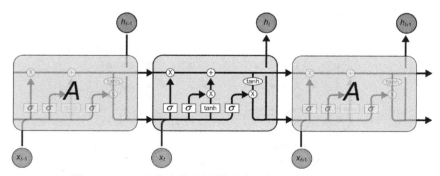

图 10.9　LSTM 网络中的重复模块包含 4 个不同的层（在 GitHub 网站中搜索"Understanding LSTM Networks"即可查看该图片）

至于其数学细节以及 4 个层之间具体是怎样交互的，读者可以在 GitHub 网站中查看具体教程。

为什么 LSTM 网络有效

通常，设计 LSTM 网络架构的第一步是决定哪部分信息是不必要的，LSTM 网络可以通过抛弃这部分信息来为更重要的信息腾出空间。为此，LSTM 网络会有一个称为**遗忘门**的层，该层会查看前一个输出 h_{t-1} 和当前输入 x_t，并决定要丢弃哪些信息。

设计 LSTM 网络架构的下一步是决定哪一部分信息是值得保留（持久地存储）在元胞中的。这一过程分两步完成。

（1）创建一个称为**输入门**的层，该层用来决定单元之前状态中的哪一些值需要更新。

（2）产生一组新的候选值，并将它们加到元胞中。

最后，我们需要决定 LSTM 元胞的输出内容。这一输出基于元胞的状态，但它还是一个过滤的版本。

10.3 语言模型的实现

本节将构建一个在字符级别操作的语言模型。为了实现这一模型，我们将用到《安娜·卡列尼娜》（*Anna Karenina*）小说，并了解网络怎么学习以实现文本的结构和风格。

该网络是基于 Andrej Karpathy 关于 RNN 的文章（在 GitHub 网站中搜索 "The Unreasonable Effectiveness of Recurrent Neural Networks" 即可查看该图片）并且使用 Torch 实现的(请在 GitHub 网站上 karpathy 的个人主页中搜索"char-rnn"）。同时，在 R2RT 网站（请在 R2RT 网站中搜索 "Recurrent Neural Networks in Tensorflow II"）上也有一些信息，还有 GitHub 网站上 Sherjil Ozairp 的信息（请在 GitHub 网站上 sherjilozair 的个人主页中搜索 "char-rnn-tensorflow"）。接下来是字符级 RNN 的常用结构。

本书将构建一个在《安娜·卡列尼娜》小说上训练的字符级 RNN，如图 10.10 所示。它能够基于书中的文本产生新的文本。读者可以在这一实现的文件中找到.txt 文件。

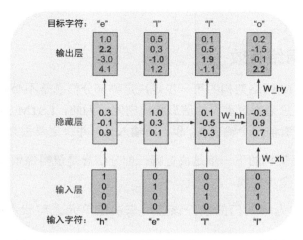

图 10.10 字符级 RNN 的通常架构（在 GitHub 网站中搜索
"The Unreasonable Effectiveness of Recurrent
Neural Networks" 即可查看该图片）

从为字符级应用导入必要的库开始。

```
import numpy as np
import tensorflow as tf

from collections import namedtuple
```

首先，需要通过加载数据集，并将它转换为整数来准备数据集。因此，将会把字符转换为整数，然后把它们作为整数来编码，这使得能够轻松地使用这些整数作为模型的输入变量。

```
#reading the Anna Karenina novel text file
with open('Anna_Karenina.txt', 'r') as f:
    textlines=f.read()

#Building the vocan and encoding the characters as integers
language_vocab = set(textlines)
vocab_to_integer = {char: j for j, char in enumerate(language_vocab)}
integer_to_vocab = dict(enumerate(language_vocab))
encoded_vocab = np.array([vocab_to_integer[char] for char in textlines],dtype=np.int32)
```

现在看看《安娜·卡列尼娜》一书中的前 200 个字符。

```
textlines[:200]
```

输出如下。

```
"Chapter 1\n\n\nHappy families are all alike; every unhappy family is unhappy
in its own\nway.\n\nEverything was in confusion in the Oblonskys' house. The
wife had\ndiscovered that the husband was carrying on"
```

我们已经把字符转换为方便网络处理的形式，也就是整数。现在看看编码以后的字符。

```
encoded_vocab[:200]
```

输出如下。

```
array([70, 34, 54, 29, 24, 19, 76, 45, 2, 79, 79, 79, 69, 54, 29, 29, 49,
       45, 66, 54, 39, 15, 44, 15, 19, 12, 45, 54, 76, 19, 45, 54, 44, 44,
       45, 54, 44, 15, 27, 19, 58, 45, 19, 30, 19, 76, 49, 45, 59, 56, 34,
       54, 29, 29, 49, 45, 66, 54, 39, 15, 44, 49, 45, 15, 12, 45, 59, 56,
       34, 54, 29, 29, 49, 45, 15, 56, 45, 15, 24, 12, 45, 11, 35, 56, 79,
       35, 54, 49, 53, 79, 79, 36, 30, 19, 76, 49, 24, 34, 15, 56, 16, 45,
       35, 54, 12, 45, 15, 56, 45, 31, 11, 56, 66, 59, 12, 15, 11, 56, 45,
       15, 56, 45, 24, 34, 19, 45, 1, 82, 44, 11, 56, 12, 27, 49, 12, 37,
       45, 34, 11, 59, 12, 19, 53, 45, 21, 34, 19, 45, 35, 15, 66, 19, 45,
       34, 54, 64, 79, 64, 15, 12, 31, 11, 30, 19, 76, 19, 64, 45, 24, 34,
       54, 24, 45, 24, 34, 19, 45, 34, 59, 12, 82, 54, 56, 64, 45, 35, 54,
       12, 45, 31, 54, 76, 76, 49, 15, 56, 16, 45, 11, 56], dtype=int32)
```

因为网络使用单个字符，所以类似于分类问题，我们可以通过之前的文本来尝试预测下一个字符。该字符的种类就是网络需要选择的类别数量。

每次往模型中输入一个字符后，模型会通过产生一个可能成为下一个字符的所有字符（词汇）的概率分布来预测下一个字符，该分布的数量相当于网络需要挑选的类别数。

```
len(language_vocab)
```

输出如下。

83

因为要用到随机梯度下降法来训练模型，所以需要把数据转换为训练批次。

10.3.1 生成训练的最小批

在本节中，我们将把数据分为用于训练小批次。所以，批次将包含许多的序列，它们具有指定数量的序列步数。下面看图 10.11 所示的一个可视化例子。

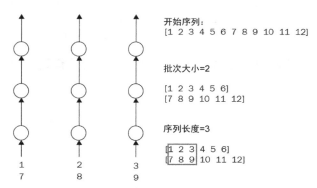

图 10.11 批次和序列的描述图（图片来源：GitHub 网站）

现在，定义一个函数来迭代地处理编码的文本并产生批次。这个函数用到了 Python 中一个很好的机制——yield（参考文章 "Improve Your Python: yield and Generators Explained"）。

一个典型的批次会包含 $N×M$ 个字符，其中，N 代表序列的个数，而 M 则是每个序列的步数。为了获得数据集可能的批次数量，可以简单地把数据的长度除以指定的批次大小，在获得可能的批次数量后，就可以知道每个批次当中应该有多少字符。

然后，需要把数据集分为指定数量（N）的序列，这可以用 arr.reshape(size) 实现。这里可以知道需要大小为 N 的序列（后面的代码中会用 num_steps 表示），令它作为数据第一维的大小。对于第二维，可以在批次大小中用–1 作为占位符，它会用合适的数据来填补数组。在这之后，就应该有大小为的 $N×$（$M×K$）的数组，其中，K 是批次的数量。

既然有了这个数组后，就可以对它进行迭代以获得批次，其中，每个批次有 $N×M$ 个字符。对于随后的批次，窗口会移动 num_steps 步。最后，还需要创建输入和输出数组来作为模型的输入。在这一步中，创建输出数值很简单，记住，标签是后移一个字符的输入。第一个输入字符通常会作为标签的最后一个字符，具体如下。

$$y[:,:-1], y[:,-1] = x[:,1:], x[:,0]$$

其中，x 是输入批次，y 是标签批次。

这里处理窗口的方法是使用 range 函数来从 0 到 arr.shape[1]（表示每个序列需要处理的总步数）取值，每次步长取 num_steps。用这种方法，使用 range 获得的整数总是指向一个批次的开头，并且每个窗口的宽度为 num_steps。

```
def generate_character_batches(data, num_seq, num_steps):
    '''Create a function that returns batches of size num_seq x num_steps
       from data.
    '''
    # Get the number of characters per batch and number of batches
    num_char_per_batch = num_seq * num_steps
    num_batches = len(data)//num_char_per_batch
    # Keep only enough characters to make full batches
    data = data[:num_batches * num_char_per_batch]
    # Reshape the array into n_seqs rows
    data = data.reshape((num_seq, -1))
    for i in range(0, data.shape[1], num_steps):
        # The input variables
        input_x = data[:, i:i+num_steps]
        # The output variables which are shifted by one
        output_y = np.zeros_like(input_x)
        output_y[:, :-1], output_y[:, -1] = input_x[:, 1:], input_x[:, 0]
        yield input_x, output_y
```

可以通过生成一个包含 15 个序列和 50 个序列步数的批次来演示这一函数。

```
generated_batches = generate_character_batches(encoded_vocab, 15, 50)
input_x, output_y = next(generated_batches)
print('input\n', input_x[:10, :10])
print('\ntarget\n', output_y[:10, :10])
```

输出如下。

```
input
 [[70 34 54 29 24 19 76 45  2 79]
 [45 19 44 15 16 15 82 44 19 45]
 [11 45 44 15 16 34 24 38 34 19]
 [45 34 54 64 45 82 19 19 56 45]
 [45 11 56 19 45 27 56 19 35 79]
 [49 19 54 76 12 45 44 54 12 24]
 [45 41 19 45 16 11 45 15 56 24]
 [11 35 45 24 11 45 39 54 27 19]
 [82 19 66 11 76 19 45 81 19 56]
 [12 54 16 19 45 44 15 27 19 45]]
```

```
target
[[34 54 29 24 19 76 45  2 79 79]
 [19 44 15 16 15 82 44 19 45 16]
 [45 44 15 16 34 24 38 34 19 54]
 [34 54 64 45 82 19 19 56 45 82]
 [11 56 19 45 27 56 19 35 79 35]
 [19 54 76 12 45 44 54 12 24 45]
 [41 19 45 16 11 45 15 56 24 11]
 [35 45 24 11 45 39 54 27 19 33]
 [19 66 11 76 19 45 81 19 56 24]
 [54 16 19 45 44 15 27 19 45 24]]
```

接下来，构建这个例子的核心——LSTM 模型。

10.3.2 构建模型

在深入讲解用 LSTM 网络构建字符级模型之前，需要先介绍一下**堆叠 LSTM（Stacked LSTM）**网络。

堆叠 LSTM 网络可用于从不同时间尺度来观察信息。

1. 堆叠 LSTM 网络

"通过将多个循环的隐藏状态相互堆叠在一起来构建深度 RNN。这种方法可能允许每个层次的隐藏状态在不同的时间尺度上操作。"见 arxiv 网站上的文章 "How to Construct Deep Recurrent Neural Networks"。

"RNN 本身在时间上就是有深度的，因为它们的隐藏状态是前面所有隐藏状态的一个函数。这篇文章提出的问题是 RNN 是否也可以从空间深度上获益，这通过相互之间堆叠多个循环的隐藏层来实现，就像前馈层堆叠在传统深度网络中一样。"见 arxiv 网站上的文章 "Speech Recognition with Deep Recurrent Neural Networks."。

大多研究者都用堆叠 LSTM 网络来解决序列预测问题。一个堆叠 LSTM 网络架构可以定义为一个包含多个 LSTM 层的 LSTM 网络模型。靠前的 LSTM 层为后续的 LSTM 层提供了一个序列输出，而不是一个单值输出。

特别是，它的每一个输入时间步对应一个输出，而不是一个输出对应所有输入时间步，如图 10.12 所示。

所以在这个例子中,我们将使用这种堆叠 LSTM 网络架构,它会有更好的性能。

2. 模型架构

网络构建在模型架构之上。本书将把模型架构分为多个部分以便逐个讲解。然后,把它们连在一起构成一个完整的网络,如图 10.13 所示。

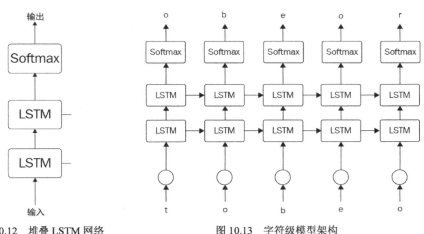

图 10.12　堆叠 LSTM 网络　　　　图 10.13　字符级模型架构

3. 输入

现在,我们从将模型输入定义为占位符开始。模型的输入是训练数据和标签。我们还会将一个名为 keep_probability 的参数用在 dropout 层中,它对防止模型过拟合会有所帮助。

```
def build_model_inputs(batch_size, num_steps):
    # Declare placeholders for the input and output variables
    inputs_x = tf.placeholder(tf.int32, [batch_size, num_steps],
name='inputs')
    targets_y = tf.placeholder(tf.int32, [batch_size, num_steps],
name='targets')
    # define the keep_probability for the dropout layer
    keep_probability = tf.placeholder(tf.float32, name='keep_prob')
    return inputs_x, targets_y, keep_probability
```

4. 构建 LSTM 元胞

在本节中,我们将写一个函数用于构造 LSTM 元胞,它将用在隐藏层中。这个元胞会成为模型的构建块，因此我们将使用 TensorFlow 来创建这个元胞。接下来看看如何使用 TensorFlow 来构建基本的 LSTM 元胞。

调用下面的一行代码来构造带有参数 num_units 的 LSTM 元胞，它表示隐藏层中的单元数量。

```
lstm_cell = tf.contrib.rnn.BasicLSTMCell(num_units)
```

为了防止过拟合，我们还要用到 dropout 技术，该技术是一种通过降低模型复杂度来防止模型在数据上过拟合的机制。

```
tf.contrib.rnn.DropoutWrapper(lstm, output_keep_prob=keep_probability)
```

正如前面提到的，这里将用到堆叠 LSTM 网络架构。堆叠 LSTM 网络架构会帮助开发人员从不同角度观察数据，并在实践中能有更好的表现。为了用 TensorFlow 定义一个堆叠 LSTM 网络，可以使用 tf.contrib.rnn.MultiRNNCell 函数（参见 TensorFlow 官网）。

```
tf.contrib.rnn.MultiRNNCell([cell]*num_layers)
```

开始，第一个元胞没有之前的信息，所以需要把元胞的状态初始为零。这可以用下面的函数来实现。

```
initial_state = cell.zero_state(batch_size, tf.float32)
```

把它们放在一起来构造一个 LSTM 元胞。

```
def build_lstm_cell(size, num_layers, batch_size, keep_probability):

    ### Building the LSTM Cell using the tensorflow function
    lstm_cell = tf.contrib.rnn.BasicLSTMCell(size)
    # Adding dropout to the layer to prevent overfitting
    drop_layer = tf.contrib.rnn.DropoutWrapper(lstm_cell, output_keep_prob=
keep_probability)
    # Add muliple cells together and stack them up to oprovide a level of more
understanding
    stakced_cell = tf.contrib.rnn.MultiRNNCell([drop_layer] * num_layers)
    initial_cell_state = lstm_cell.zero_state(batch_size, tf.float32)
    return lstm_cell, initial_cell_state
```

5. RNN 输出

接下来，需要构造输出层。输出层需要读取单个 LSTM 元胞的输出，然后把它们传递到全连接层。输出层有一个 softmax 输出，用来产生输入字符之后可能的下一个字符的概率分布。

之前已经为网络生成了大小为 N×M 个字符的输入批次，其中，N 是批次中的序列数量，而 M 是序列的步数。在创建模型时，我们还在隐藏层中用了 L 个隐藏单元。基于批次大小以及隐藏单元个数，网络的输出将是一个 N×M×L 的三维张量，这是因为调用了 LSTM 元胞 M 次，每一个序列步骤调用一次。每一次调用 LSTM 元胞都会产生一个大小为 L 的输出。最后，需要重复的次数等于拥有的序列数量 N。

把 N×M×L 的输出输入全连接层里（对所有的输出都有一样的权重）。但在做这个之前，需要把输出调整为 2 维张量，它的形状是（M×N）×L。这个调整会使得在操作输出时更加方便，因为新的形状会有更多的便利：每一列都表示 LSTM 元胞的 L 个输出，因此这是每一个序列的每一步里的一列。

得到新的形状后，可以通过和权重进行矩阵相乘，从而与带 softmax 的全连接层相连。在 LSTM 元胞里已构建的权重和这里将创建的权重默认有相同的名字，在这种情况下 TensorFlow 会报错。为了避免这个错误，可以用 TensorFlow 的函数 tf.variable_scope() 将这里创建的权重和偏置变量放在一个变量作用域内。

在解释输出的形状以及如何重塑它之前，为了让事情变得简单一点，先定义一个辅助函数 build_model_output。

```
def build_model_output(output, input_size, output_size):

    # Reshaping output of the model to become a bunch of rows, where each row
correspond for each step in the seq
    sequence_output = tf.concat(output, axis=1)
    reshaped_output = tf.reshape(sequence_output, [-1, input_size])
    # Connect the RNN outputs to a softmax layer
    with tf.variable_scope('softmax'):
        softmax_w = tf.Variable(tf.truncated_normal((input_size, output_size),
stddev=0.1))
        softmax_b = tf.Variable(tf.zeros(output_size))
    # the output is a set of rows of LSTM cell outputs, so the logits will be a set
    # of rows of logit outputs, one for each step and sequence
    logits = tf.matmul(reshaped_output, softmax_w) + softmax_b
    # Use softmax to get the probabilities for predicted characters
    model_out = tf.nn.softmax(logits, name='predictions')
    return model_out, logits
```

6. 训练损失

接下来，训练损失。我们可以获得对数和标签并计算 softmax 交叉熵。首先，需要对标签进行独热编码，以编码字符的格式获取它们。然后，调整独热标签的形状，使它成为一个（M*N）×C 大小的二维向量，其中，C 是所拥有的类别（字符）数量。之前我们调整了 LSTM 层输出的形状，并让它们通过带有 C 个单元的全连接层。所以，对数形式的大小也为（M*N）×C。

然后，用 tf.nn.softmax_cross_entropy_with_logits 来运行 logits 和 targets，并计算平均值以获得损失。

```
def model_loss(logits, targets, lstm_size, num_classes):
    # convert the targets to one-hot encoded and reshape them to match the logits, one row per batch_size per step
    output_y_one_hot = tf.one_hot(targets, num_classes)
    output_y_reshaped = tf.reshape(output_y_one_hot, logits.get_shape())
    #Use the cross entropy loss
    model_loss = tf.nn.softmax_cross_entropy_with_logits(logits=logits, labels=output_y_reshaped)
    model_loss = tf.reduce_mean(model_loss)
    return model_loss
```

7. 优化器

最后，我们需要用一个优化方法以从数据集中学习东西。之前讲到，普通 RNN 存在梯度爆炸和梯度消失问题。LSTM 网络只解决了其中一个问题，即梯度值消失问题。但即使在使用了 LSTM 网络后，也存在部分梯度值爆炸且无限增长的可能性。为了解决这个问题，可以用**梯度裁剪法**，这个方法会把超过某个特定阈值的梯度裁剪掉。

现在可以通过用 Adam 优化器为学习这个过程定义优化器。

```
def build_model_optimizer(model_loss, learning_rate, grad_clip):

    # define optimizer for training, using gradient clipping to avoid the exploding of the gradients
    trainable_variables = tf.trainable_variables()
    gradients, _ = tf.clip_by_global_norm(tf.gradients(model_loss, trainable_variables), grad_clip)
    #Use Adam Optimizer
```

```
        train_operation = tf.train.AdamOptimizer(learning_rate)
        model_optimizer = train_operation.apply_gradients(zip(gradients, trainable_
variables))
        return model_optimizer
```

8. 构建网络

现在，我们可以把所有的片段连在一起以构建一个网络的类。为了切实在 LSTM 元胞上运行数据，将会用到 tf.nn.dynamic_rnn（见 TensorFlow 官网）。这个函数会把隐藏的状态和元胞状态通过 LSTM 元胞合理地传递过来。它会返回最小批次里每个序列每一步的每一个 LSTM 元胞输出，同时它还会给出最后的 LSTM 状态。我们想把这个状态另存为 final_state，从而可以把它传递给下一个最小批次的第一个 LSTM 元胞。对于 tf.nn.dynamic_rnn，我们向它传入从 build_lstm 获得的元胞、起始状态以及输入序列。同时，在将输入送入 RNN 之前，需要对输入进行独热编码。

```
class CharLSTM:
    def __init__(self, num_classes, batch_size=64, num_steps=50,
                 lstm_size=128, num_layers=2, learning_rate=0.001,
                 grad_clip=5, sampling=False):
        # When we're using this network for generating text by sampling, we'll be
providing the network with
        # one character at a time, so providing an option for it.
        if sampling == True:
            batch_size, num_steps = 1, 1
        else:
            batch_size, num_steps = batch_size, num_steps

        tf.reset_default_graph()
        # Build the model inputs placeholders of the input and target variables
        self.inputs, self.targets, self.keep_prob = build_model_inputs(batch_
size, num_steps)

        # Building the LSTM cell
        lstm_cell, self.initial_state = build_lstm_cell(lstm_size, num_layers,
batch_size, self.keep_prob)

        # Run the data through the LSTM layers
        # one_hot encode the input
        input_x_one_hot = tf.one_hot(self.inputs, num_classes)
        # Runing each sequence step through the LSTM architecture and finally
collecting the outputs
```

```
        outputs, state = tf.nn.dynamic_rnn(lstm_cell, input_x_one_hot, initial_
state=self.initial_state)
        self.final_state = state
        # Get softmax predictions and logits
        self.prediction, self.logits = build_model_output(outputs, lstm_size,
num_classes)
        # Loss and optimizer (with gradient clipping)
        self.loss = model_loss(self.logits, self.targets, lstm_size, num_classes)
        self.optimizer = build_model_optimizer(self.loss, learning_rate, grad_clip)
```

9. 模型超参数

与其他深度学习架构一样，有一些超参数可用来控制模型并对它进行微调。以下是用于此架构的超参数集。

- 批次大小是每次输入网络里的序列数量。
- 步数是训练网络的序列中的字符数，通常，步数越大越好，以便网络可以学到更多的长期依赖关系，但是这会导致更长的训练时间。在这里，100 一般是一个比较好的步数值。
- LSTM 大小是隐藏层的单元数量。
- 架构层数是要用到的隐藏 LSTM 层数。
- 学习率是通常的训练学习率。
- 最后，名为保留概率的新参数是在 dropout 层中使用的，它有助于网络避免过拟合。所以如果网络过拟合了，可以尝试减小这个值。

10.3.3 训练模型

现在，我们通过向构建的模型提供输入和输出，然后使用优化器来训练网络。在为当前状态做预测时，不要忘记我们需要用到之前的状态。因此，我们需要将输出状态传递回网络，以便在预测下一个输入时使用它。

现在为超参数提供初始值（可以根据用于训练此架构的数据集来调整它们）。

```
batch_size = 100        # Sequences per batch
num_steps = 100         # Number of sequence steps per batch
lstm_size = 512         # Size of hidden layers in LSTMs
num_layers = 2          # Number of LSTM layers
```

10.3 语言模型的实现

```python
learning_rate = 0.001    # Learning rate
keep_probability = 0.5   # Dropout keep probability

epochs = 5

# Save a checkpoint N iterations
save_every_n = 100

LSTM_model = CharLSTM(len(language_vocab), batch_size=batch_size,
num_steps=num_steps,
                lstm_size=lstm_size, num_layers=num_layers,
                learning_rate=learning_rate)

saver = tf.train.Saver(max_to_keep=100)
with tf.Session() as sess:
    sess.run(tf.global_variables_initializer())
    # Use the line below to load a checkpoint and resume training
    # saver.restore(sess, 'checkpoints/_____.ckpt')
    counter = 0
    for e in range(epochs):
        # Train network
        new_state = sess.run(LSTM_model.initial_state)
        loss = 0
        for x, y in generate_character_batches(encoded_vocab, batch_size, num_steps):
            counter += 1
            start = time.time()
            feed = {LSTM_model.inputs: x,
                    LSTM_model.targets: y,
                    LSTM_model.keep_prob: keep_probability,
                    LSTM_model.initial_state: new_state}
            batch_loss, new_state, _ = sess.run([LSTM_model.loss,
                                                LSTM_model.final_state,
                                                LSTM_model.optimizer],
                                                feed_dict=feed)
            end = time.time()
            print('Epoch number: {}/{}... '.format(e+1, epochs),
                  'Step: {}... '.format(counter),
                  'loss: {:.4f}... '.format(batch_loss),
                  '{:.3f} sec/batch'.format((end-start)))
            if (counter % save_every_n == 0):
                saver.save(sess, "checkpoints/i{}_l{}.ckpt".format(counter, lstm_size))
    saver.save(sess, "checkpoints/i{}_l{}.ckpt".format(counter, lstm_size))
```

在训练过程结束时，应该会得到和以下值接近的误差。

.
.
.

```
Epoch number: 5/5... Step: 978... loss: 1.7151... 0.050 sec/batch
Epoch number: 5/5... Step: 979... loss: 1.7428... 0.051 sec/batch
Epoch number: 5/5... Step: 980... loss: 1.7151... 0.050 sec/batch
Epoch number: 5/5... Step: 981... loss: 1.7236... 0.050 sec/batch
Epoch number: 5/5... Step: 982... loss: 1.7314... 0.051 sec/batch
Epoch number: 5/5... Step: 983... loss: 1.7369... 0.051 sec/batch
Epoch number: 5/5... Step: 984... loss: 1.7075... 0.065 sec/batch
Epoch number: 5/5... Step: 985... loss: 1.7304... 0.051 sec/batch
Epoch number: 5/5... Step: 986... loss: 1.7128... 0.049 sec/batch
Epoch number: 5/5... Step: 987... loss: 1.7107... 0.051 sec/batch
Epoch number: 5/5... Step: 988... loss: 1.7351... 0.051 sec/batch
Epoch number: 5/5... Step: 989... loss: 1.7260... 0.049 sec/batch
Epoch number: 5/5... Step: 990... loss: 1.7144... 0.051 sec/batch
```

1. 保存检查点

现在，开始加载检查点。如果要了解更多保存和加载检查点的信息，请查看相关 TensorFlow 文档（见 TensorFlow 官网）。

```
tf.train.get_checkpoint_state('checkpoints')
```

输出如下。

```
model_checkpoint_path: "checkpoints/i990_l512.ckpt"
all_model_checkpoint_paths: "checkpoints/i100_l512.ckpt"
all_model_checkpoint_paths: "checkpoints/i200_l512.ckpt"
all_model_checkpoint_paths: "checkpoints/i300_l512.ckpt"
all_model_checkpoint_paths: "checkpoints/i400_l512.ckpt"
all_model_checkpoint_paths: "checkpoints/i500_l512.ckpt"
all_model_checkpoint_paths: "checkpoints/i600_l512.ckpt"
all_model_checkpoint_paths: "checkpoints/i700_l512.ckpt"
all_model_checkpoint_paths: "checkpoints/i800_l512.ckpt"
all_model_checkpoint_paths: "checkpoints/i900_l512.ckpt"
all_model_checkpoint_paths: "checkpoints/i990_l512.ckpt"
```

2. 生成文本

现在我们已经有了基于输入数据集的训练好的模型。下一步是用这个训练好的模型来生成文本,并看看这个模型如何从输入数据中学到风格和结构。为了达到这个目的,我们可以从一些初始字符开始,然后以新的、预测出来的字符作为下一步的输入。这个步骤将一直重复,直至得到一个指定长度的文本。

在接下来的代码中,我们还在函数中加入了一些额外的语句,来用一些初始文本启动网络,并从那里开始运行。

网络将会给出预测结果或者词汇表里每个字符的概率。为了降低噪声并只用网络比较确定的字符,我们将只从输出中最可能的 N 个字符中挑选一个新字符。

```
def choose_top_n_characters(preds, vocab_size, top_n_chars=4):
    p = np.squeeze(preds)
    p[np.argsort(p)[:-top_n_chars]] = 0
    p = p / np.sum(p)
    c = np.random.choice(vocab_size, 1, p=p)[0]
    return c

def sample_from_LSTM_output(checkpoint, n_samples, lstm_size, vocab_size, prime=
"The "):
    samples = [char for char in prime]
    LSTM_model = CharLSTM(len(language_vocab), lstm_size=lstm_size, sampling=
True)
    saver = tf.train.Saver()
    with tf.Session() as sess:
        saver.restore(sess, checkpoint)
        new_state = sess.run(LSTM_model.initial_state)
        for char in prime:
            x = np.zeros((1, 1))
            x[0,0] = vocab_to_integer[char]
            feed = {LSTM_model.inputs: x,
                    LSTM_model.keep_prob: 1.,
                    LSTM_model.initial_state: new_state}
            preds, new_state = sess.run([LSTM_model.prediction,
LSTM_model.final_state],
                                        feed_dict=feed)

        c = choose_top_n_characters(preds, len(language_vocab))
        samples.append(integer_to_vocab[c])

        for i in range(n_samples):
            x[0,0] = c
            feed = {LSTM_model.inputs: x,
```

```
                        LSTM_model.keep_prob: 1.,
                        LSTM_model.initial_state: new_state}
                preds, new_state = sess.run([LSTM_model.prediction,
LSTM_model.final_state],
                                            feed_dict=feed)

                c = choose_top_n_characters(preds, len(language_vocab))
                samples.append(integer_to_vocab[c])
        return ''.join(samples)
```

用最新保存的检查点来开始采样过程。

```
tf.train.latest_checkpoint('checkpoints')
```

输出如下。

```
'checkpoints/i990_l512.ckpt'
```

下面开始用最新的检查点来采样：

```
checkpoint = tf.train.latest_checkpoint('checkpoints')
sampled_text = sample_from_LSTM_output(checkpoint, 1000, lstm_size,
len(language_vocab), prime="Far")
print(sampled_text)
```

输出如下。

```
INFO:tensorflow:Restoring parameters from checkpoints/i990_l512.ckpt

Farcial the
confiring to the mone of the correm and thinds. She
she saw the
streads of herself hand only astended of the carres to her his some of the
princess of which he came him of
all that his white the dreasing of
thisking the princess and with she was she had
bettee a still and he was happined, with the pood on the mush to the
peaters and seet it.

"The possess a streatich, the may were notine at his mate a misted
and the
man of the mother at the same of the seem her
felt. He had not here.
```

```
"I conest only be alw you thinking that the partion
of their said."

"A much then you make all her
somether. Hower their centing
about
this, and I won't give it in
himself.
I had not come at any see it will that there she chile no one that him.

"The distiction with you all.... It was
a mone of the mind were starding to the simple to a mone. It to be to ser
in the place," said Vronsky.
"And a plais in
his face, has alled in the consess on at they to gan in the sint
at as that
he would not be and t
```

可以看到，模型能够生成一些有意义的单词和一些无意义的单词。为了获得更多结果，我们可以运行模型更长的时间，并尝试调整超参数。

10.4　总结

本章学习了 RNN 及其工作原理，以及为什么它们变得如此重要。我们在有趣的小说数据集上训练了一个 RNN 字符级语言模型，并看到了 RNN 的结果。值得期待的是，RNN 会有更多的创新，我也相信 RNN 将成为智能系统中普遍且关键的组成部分。

第 11 章
表示学习——实现词嵌入

机器学习是一门主要基于统计学和线性代数的科学。在多数机器学习或由于反向传播的深度学习架构中，运用矩阵运算是十分常见的。这是深度学习或者通常的机器学习只接受实数值作为输入的主要原因。这个情况与很多应用都相互矛盾，如机器翻译、情感分析等，它们都以文本作为输入。所以为了在这些应用中用到深度学习，我们需要把输入转换为深度学习接受的格式。

本章将介绍表示学习这一领域，它是从文本中学习一个实数表示同时保留真实文本的语义信息的一种方法。例如，love 的表示应该和 adore 的表示十分接近，因为它们的应用情景十分相似。

本章包含以下主题。

- 表示学习简介。
- Word2Vec。
- skip-gram 架构的一个实际例子。
- 实现 skip-gram Word2Vec。

11.1　表示学习简介

到目前为止，本书讲到的所有机器学习算法或架构都要求输入是实数或者实数矩阵，这是机器学习共有的一个主题。例如，在卷积神经网络里，我们需要以输入图像的原始像素值作为模型输入。在这一部分，我们将要处理文本，因此需要用某种方式编码文本，并产生可以输入机器学习算法里的实数值。为了把输入文本编码为实数值，需要用到一门中间学科,该学科叫作**自然语言处理**(Natural Language Processing,NLP)。

之前提到过像情感分析这种给机器学习模型提供文本的流程,这可能是有问题且不可行的,因为不能把反向传播或者其他诸如点积的运算应用在输入上,原因在于它是字符串。因此,我们需要 NLP 机制的帮助,它将帮助构建文本的中间表示。该中间表示携带和文本一样的信息,并且可以输入机器学习模型中。

我们需要把输入文本里的每个单词或标志转换为一个实数向量。这些向量如果不携带原始输入的模型、信息、意义和语义,那么它们将是没有用处的。例如,在真实的文本中,单词 love 和 adore 是十分相似的,并具有相同的意思。开发人员需要能够表示它们之间距离很近并且位于同一向量空间中的实值向量。因此,这两个单词以及与它们不相似的另一个单词的向量表示如图 11.1 所示。

有很多技术可以用于完成这项任务,这类技术称为**嵌入**(**embedding**),这类技术把文本嵌入另一个实数向量空间中。

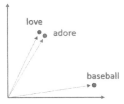

正如后续将看到的,向量空间事实上是十分有趣的,因为你会发现,可以从其他与之类似的单词中推导出一个单词的向量表示,甚至可以在这个空间上做一些几何操作。

图 11.1 单词的向量表示

11.2 Word2Vec

Word2Vec 是 NLP 领域里常用的嵌入技术之一。这个模型会通过查看输入单词出现的上下文信息来从输入中构建实数向量。因此我们会发现相似的单词会在相似的上下文中提到,因此模型会学到,那两个单词在特定的嵌入空间里应该距离很近。

从图 11.2 中的语句,模型会学到:单词 **love** 和单词 **adore** 有相似的上下文,并且在最终的向量空间里应该距离很近。like 的上下文可能与 love 类似,但是它不像接近 love 那样接近 adore。

I love taking long walks on the beach.
My friends told me that they love popcorn.

The relatives adore the baby's cute face.
I adore his sense of humor.

图 11.2 情感语句的样本

Word2Vec 模型也依赖输入句子的语义特征。例如,adore 和 love 两个单词都主要用在积极的上下文中,并且通常位于名词或名词短语前。同时,模型也会学到这两个单词有些地方相似,并且它很有可能将这两个单词的向量表示放在相似的上下文中。所以,句子的结构能够透露给 Word2Vec 模型很多关于相似单词的信息。

在实践中,开发人员把大型的文本语料库输入 Word2Vec 模型。模型会学习为相似的单词产生相似的向量,并且对文本中的每一个单词都会这么做。

所有这些单词的向量会合并在一起,最后的输出会是一个嵌入矩阵,其中,每一行表示一个特定单词的实数向量表示。Word2Vec 模型流程的示例如图 11.3 所示。

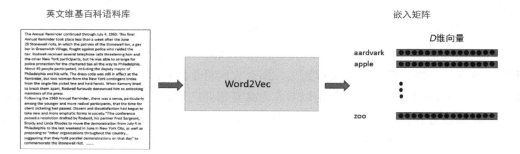

图 11.3　Word2Vec 模型流程的例子

模型最后的输出是一个训练语料库中所有单词的嵌入矩阵。通常,好的嵌入矩阵可以包含上百万个实数向量。

Word2Vec 模型用一个窗口来扫描句子,然后根据上下文信息预测窗口中间的单词的向量。Word2Vec 每次会扫描一个句子。和所有的机器学习技术相似,我们需要为 Word2Vec 模型定义一个损失函数和它对应的优化准则,它们将使得模型能够为每个单词产生实数向量,并且基于单词的上下文信息把向量互相关联。

构建 Word2Vec 模型

本节将讨论关于如何构建 Word2Vec 模型的一些深层细节。正如前面提到的,最后的目标是有一个训练好的模型,它能够为输入的文本数据产生实数向量表示,这也称为词嵌入。

在模型训练过程中,我们会用到最大似然法(参见维基百科),给定模型前面见过的单词,这里用 h 表示,然后最大化输入句子中下一个单词 w_t 的概率。

这个最大似然法可以以 Softmax 函数的形式给出。

$$P(w_t / h) = \text{Softmax}(\text{score}(w_t, h))$$
$$= \frac{\exp\{\text{score}(w_t, h)\}}{\sum_{\text{Word } w' \text{ in Vocab}} \exp\{\text{score}(w', h)\}}$$

其中,score 函数计算了一个值,它表示目标单词 w_t 和上下文 h 的兼容性。会在输入句子上训练该模型,训练的目的是最大化这些训练输入数据上的似然(使用对数似然是为了数学上的简便,所以使用了对数来推导)。

$$J_{\text{ML}} = \log P(w_t \mid h)$$
$$= \text{score}(w_t, h) - \log\left(\sum_{\text{Word } w' \text{ in Vocab}} \exp\{\text{score}(w', h)\}\right)$$

所以,ML(最大似然)法会尝试最大化上面的等式,从而得到一个概率语言模型,如图11.4所示。这里的计算是十分复杂的,因为需要用score函数计算词汇表V中的每一个单词w'在该模型对应的当前上下文h中的概率。这种计算在每一个训练步骤里都要进行。

因为构建概率语言模型所需的计算非常复杂,所以开发人员偏向于用计算相对不复杂的其他技术来实现,如**连续词袋(Continuous Bag-of-Words,CBOW)**和skip-gram模型。

训练这些模型以构建一个带logistic回归的二分类任务,用于区分真实的目标单词w_t、h噪声或者假想单词$\tilde{w}$,这些是在同一上下文中完成的。图11.5用CBOW技术简化了这个想法。

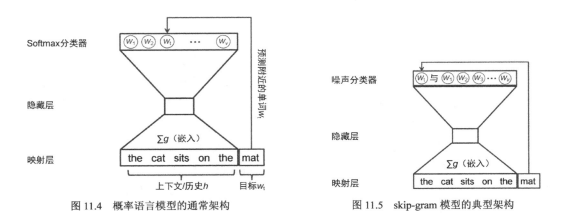

图11.4 概率语言模型的通常架构　　　图11.5 skip-gram模型的典型架构

图11.6展示了可以用于构架Word2Vec模型的两个架构。

更加具体地说,这些技术的目标是最大化下面的式子。

$$J_{\text{NEG}} = \log Q_\theta(D=1 \mid w_t, h) + k \mathop{\mathbb{E}}_{\tilde{w} \sim P_{\text{noise}}}[\log Q_\theta(D=0 \mid \tilde{w}, h)]$$

其中,$Q_\theta(D=1 \mid w_t, h)$是模型基于单词$w$和数据集$D$中上下文$h$的二值logistic回

归概率，它是用 θ 向量计算的，这个向量代表所学的嵌入；$\tilde{w}$ 是假想单词或者噪声单词，它可以从一个噪声概率分布中产生，例如，训练的输入例子的一元分布。

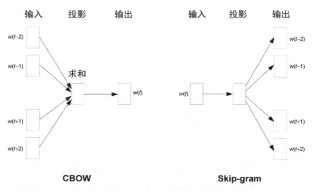

图 11.6　Word2Vec 模型的不同架构

总的来说，这些模型的目标就是区分真实输入和假想输入，因此会为真实单词分配较高的概率，而为假想单词或者噪声单词分配较低的概率。

如果模型为真实单词分配较高的概率，而为噪声单词分配较低的概率，目标函数就最大化了。

> 从技术上讲，为真实单词分配高概率的过程称为**负抽样**（**negative sampling**，详见 nips 网站）。使用此损失函数有良好的数学动机：它提出的更新近似于限制条件中 softmax 函数的更新。从计算上讲，它特别有吸引力，因为计算损失函数现在只能根据选择的噪声单词的数量（k）来缩放，而不是词汇表中的所有词（V）。这使得训练更快。实际上，这里将会使用十分相似的**噪声对比估计**（**noise-contrastive estimation**，**NCE**）损失（关于 NCE，详见 nips 网站），它在 TensorFlow 里有一个十分方便的辅助函数——tf.nn.nce_loss()。

11.3　skip-gram 架构的一个实际例子

现在介绍一个实际的例子，并看看在这种情况下 skip-gram 模型是如何工作的。

```
the quick brown fox jumped over the lazy dog
```

11.3 skip-gram 架构的一个实际例子

首先，需要构建一个包含单词和对应上下文的数据集。怎样定义背景取决于开发人员，但必须合理。我们将在目标单词周围放一个窗口，并且从单词的右边和左边各取一个单词。

通过这种上下文技术，我们最后会得到下列单词和对应上下文的集合。

([the, brown], quick), ([quick, fox], brown), ([brown, jumped], fox), ...

生成的词语和它们对应的上下文会用一对（context, target）表示。skip-gram 模型的想法和 CBOW 的想法相反。skip-gram 模型将会根据目标单词预测它的背景。例如，考虑上面的第一对，skip-gram 模型会尝试从目标单词 quick 预测 the 和 brown 等。所以数据集也可以重写成如下形式。

(quick, the), (quick, brown), (brown, quick), (brown, fox), ...

现在，我们已经有了输入和输出对。

下面可以尝试模仿第 t 步的训练过程。skip-gram 模型会接受第一个训练样本，其中，输入是单词 quick，而目标输出是单词 the。接着，我们还需要构造噪声输入，并随机地从输入数据的一元模型中选择一个单词。为了方便，噪声向量的大小取为 1。例如，可以选择单词 sheep 作为噪声。

现在，继续计算真实配对和噪声配对的损失。

$$J_{\text{NEG}}^{(t)} = \log Q_\theta(D=1 \mid \text{the, quick}) + \log(Q_\theta(D=0 \mid \text{sheep, quick}))$$

这里的目标是更新 θ 参数来改善之前的目标函数。通常，可以使用梯度来做这件事，因此需要计算损失关于目标函数参数 θ 的梯度，这个梯度将会表示为 $\frac{\partial}{\partial \theta} J_{\text{NEG}}$。

在训练过程后，可以基于降维后的实数向量表示方式来可视化结果。读者将会发现这个向量空间十分有趣，我们可以使用它做很多有趣的事情。例如，可以在这个空间里做类比，如 king 之于 queen 就像 man 之于 woman，甚至还可以通过从 queen 向量减去 king 向量再加上 man 向量得到 woman 向量，这个结果将会和实际学习的 woman 向量十分接近，如图 11.7 所示。读者还可以在这个空间里学习几何知识。

前面的例子给出了这些向量背后很好且直观的感受，以及它们是如何在类似机器翻译或**词性**（part-of-speech，POS）标注等 NLP 应用里起到作用。

男性-女性　　　　　时态　　　　　国家首都

图11.7　用t分布式随机邻域嵌入（t-SNE）降维法可视化学习的向量在二维空间上的投影

11.4　实现 skip-gram Word2Vec

在了解了 skip-gram 模型如何工作的数学细节之后，接下来我们实现 skip-gram。skip-gram 把词语编码成带有某种性质的实数向量（因此称为 Word2Vec）。通过实现这个架构，我们将得到学习另一种表示的过程如何工作的线索。

文本是许多类似机器翻译、情感分析、文本语音转换系统等自然语言处理应用的主要输入。因此，为文本学习一个实数表示可以帮助开发人员用不同的深度学习技术解决任务。

前面介绍了名为独热编码的技术，该技术产生一个除了这个向量所表示单词的索引外其余索引全为零的向量。为什么这里不用它呢？因为独热编码是一个十分低效的技术，通常开发人员有大量不同的单词，例如，可能有 50 000 个单词，那么使用独热向量就会产生一个其中有 49999 个分量为 0、1 个分量为 1 的向量。

这样一个稀疏的输入会导致巨大的计算浪费，原因是我们会在神经网络隐藏层中做矩阵乘法，如图 11.8 所示。

正如前面所提到的，使用独热编码会生成一个十分稀疏的向量，特别是当拥有大量要编码的不同单词的时候。

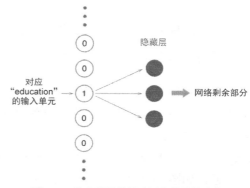

图11.8　独热编码将导致巨大的计算浪费

如图 11.9 所示，当使用这个独热向量乘以一个权重矩阵后，输出就会仅仅是权重矩阵中的一行，该行就是独热向量中不为零的那一个值与权重矩阵第 4 行相乘的结果。

为了避免这种巨大的计算浪费，我们将会使用嵌入技术，也就是一个带嵌入权重的全连接层。在这一层当中，我们会避免低效的乘法并从名为**权重矩阵**的地方开始查找嵌入层的嵌入权重。

比起因计算产生的浪费，我们更倾向于使用权重来查找权重矩阵从而得到嵌入权重。首先，需要实现这个查找。为此，要把所有的输入单词编码为整数，如图 11.10 所示。然后，找到这个单词的对应值，这里将它的整数表示作为权重矩阵里的行数。寻找某个特定词语的对应嵌入值的过程称为**嵌入查找（embedding lookup）**。正如前面所提到的，嵌入层会是一个全连接层，其中，单元的个数表示嵌入维度。

$$\begin{bmatrix}0 & 0 & 0 & 1 & 0\end{bmatrix} \times \begin{bmatrix}8 & 2 & 1 & 9\\ 6 & 5 & 4 & 0\\ 7 & 1 & 6 & 2\\ 1 & 3 & 5 & 8\\ 0 & 4 & 9 & 1\end{bmatrix} = \begin{bmatrix}1 & 3 & 5 & 8\end{bmatrix}$$

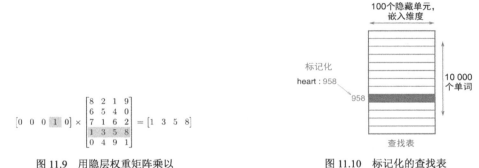

图 11.9 用隐层权重矩阵乘以几乎全为零的独热向量的结果

图 11.10 标记化的查找表

可以看到，这个过程是非常直观且简单的，只需要依照以下步骤进行即可。

（1）定义查找表，并将它看作一个权重矩阵。

（2）将嵌入层定义为一个全连接隐藏层，该层带有特定数量的单元（嵌入维度）。

（3）使用权重矩阵查找来替代不需要计算的矩阵乘法。

（4）像所有权重矩阵一样训练查找表。

正如前面提到的，本节将构建一个 skip-gram Word2Vec 模型，它是学习单词表示的有效方法，同时还能保留单词拥有的语义信息。

现在，我们继续用 skip-gram 架构来构建一个 Word2Vec 模型，已证明它比其他的架构效果更好。

11.4.1 数据分析与预处理

本节将定义一些辅助函数，它们有助于构建一个好的 Word2Vec 模型。为了实现这个目标，我们将会用到一个净化版的维基百科（详见 mattmahoney 网站）。

现在，从为该实现导入所需要的库开始。

```python
#importing the required packages for this implementation
import numpy as np
import tensorflow as tf

#Packages for downloading the dataset
from urllib.request import urlretrieve
from os.path import isfile, isdir
from tqdm import tqdm
import zipfile

#packages for data preprocessing
import re
from collections import Counter
import random
```

接下来，定义一个类，该类用来下载数据集（如果你之前没有下载过它）。

```python
# In this implementation we will use a cleaned up version of Wikipedia from Matt Mahoney.
# So we will define a helper class that will helps to download the dataset
wiki_dataset_folder_path = 'wikipedia_data'
wiki_dataset_filename = 'text8.zip'
wiki_dataset_name = 'Text8 Dataset'

class DLProgress(tqdm):
    last_block = 0

    def hook(self, block_num=1, block_size=1, total_size=None):
        self.total = total_size
        self.update((block_num - self.last_block) * block_size)
        self.last_block = block_num
# Cheking if the file is not already downloaded
if not isfile(wiki_dataset_filename):
    with DLProgress(unit='B', unit_scale=True, miniters=1, desc=wiki_dataset_name) as pbar:
        urlretrieve(
            'http://mattmahoney.net/dc/text8.zip',
            wiki_dataset_filename,
            pbar.hook)
```

11.4 实现 skip-gram Word2Vec

```
# Checking if the data is already extracted if not extract it
if not isdir(wiki_dataset_folder_path):
    with zipfile.ZipFile(wiki_dataset_filename) as zip_ref:
        zip_ref.extractall(wiki_dataset_folder_path)
with open('wikipedia_data/text8') as f:
    cleaned_wikipedia_text = f.read()
```

输出如下。

```
Text8 Dataset: 31.4MB [00:39, 794kB/s]
```

可以看看这个数据集的前 100 个字符。

```
cleaned_wikipedia_text[0:100]
```

```
' anarchism originated as a term of abuse first used against early working class radicals including t'
```

接下来,我们将对文本进行预处理,因此需要定义辅助函数。该辅助函数将有助于把诸如标点符号等特殊字符替换为已知的标记。同时,为了降低输入文本中的噪声,开发人员可能想移除文本中不常出现的单词。

```
def preprocess_text(input_text):

    # Replace punctuation with some special tokens so we can use them in our model
    input_text = input_text.lower()
    input_text = input_text.replace('.', ' <PERIOD> ')
    input_text = input_text.replace(',', ' <COMMA> ')
    input_text = input_text.replace('"', ' <QUOTATION_MARK> ')
    input_text = input_text.replace(';', ' <SEMICOLON> ')
    input_text = input_text.replace('!', ' <EXCLAMATION_MARK> ')
    input_text = input_text.replace('?', ' <QUESTION_MARK> ')
    input_text = input_text.replace('(', ' <LEFT_PAREN> ')
    input_text = input_text.replace(')', ' <RIGHT_PAREN> ')
    input_text = input_text.replace('--', ' <HYPHENS> ')
    input_text = input_text.replace('?', ' <QUESTION_MARK> ')
    input_text = input_text.replace(':', ' <COLON> ')
    text_words = input_text.split()
    # neglecting all the words that have five occurrences of fewer
    text_word_counts = Counter(text_words)
```

```
    trimmed_words = [word for word in text_words if text_word_counts
[word] > 5]

    return trimmed_words
```

在输入文本上调用这个函数，并查看输出结果。

```
preprocessed_words = preprocess_text(cleaned_wikipedia_text)
print(preprocessed_words[:30])
```

输出如下。

```
['anarchism', 'originated', 'as', 'a', 'term', 'of', 'abuse', 'first',
'used', 'against', 'early', 'working', 'class', 'radicals', 'including',
'the', 'diggers', 'of', 'the', 'english', 'revolution', 'and', 'the',
'sans', 'culottes', 'of', 'the', 'french', 'revolution', 'whilst']
```

下面查看预处理版本的文本中有多少单词和多少不同的单词。

```
print("Total number of words in the text:
{}".format(len(preprocessed_words)))
print("Total number of unique words in the text:
{}".format(len(set(preprocessed_words))))
```

输出如下。

```
Total number of words in the text: 16680599
Total number of unique words in the text: 63641
```

在这里，通过创建字典来将单词转换为整数，相反，也将整数转换为单词。整数按频率降序分配，因此把整数 0 分配给最常用的单词（the），把 1 分配给下一个最常用的单词，依次类推。把单词转换为整数并存储在列表 int_words 中。

如本节前面所述，读者需要使用单词的整数索引在权重矩阵中查找它们的值，因此将单词转换为整数，同时将整数转换为单词。这将有助于读者查找单词，同时也可以获得特定索引处的确切单词。例如，输入文本中重复次数最多的单词将在位置 0 处索引，然后是重复次数第二多的单词，依次类推。

因此，这里定义一个函数来创建这个查找表。

```
def create_lookuptables(input_words):
    """
    Creating lookup tables for vocan
```

```
Function arguments:
param words: Input list of words
"""
input_word_counts = Counter(input_words)
sorted_vocab = sorted(input_word_counts, key=input_word_counts.get, reverse=True)
integer_to_vocab = {ii: word for ii, word in enumerate(sorted_vocab)}
vocab_to_integer = {word: ii for ii, word in integer_to_vocab.items()}

# returning A tuple of dicts
return vocab_to_integer, integer_to_vocab
```

现在，调用定义的函数来创建查找表。

```
vocab_to_integer, integer_to_vocab = create_lookuptables(preprocessed_words)
integer_words = [vocab_to_integer[word] for word in preprocessed_words]
```

为了构建一个更加精确的模型，可以移除对上下文影响不大的单词，如 of、for、the 等。该行为实际上证明了开发人员可以在丢弃这类单词的同时构建更准确的模型。从上下文中删除与上下文无关的单词的过程称为**子采样**。为了定义单词丢弃的一般机制，Mikolov 引入了一个函数来计算某个单词的丢弃概率，具体如下。

$$P(w_i) = 1 - \sqrt{\frac{t}{f(w_i)}}$$

其中，t 是丢弃单词的阈值参数；$f(w_i)$ 是输入数据集中某个特定目标单词 w_i 的频率。

接下来，我们将实现一个辅助函数来计算数据集中每个单词的丢弃概率。

```
# removing context-irrelevant words threshold
word_threshold = 1e-5

word_counts = Counter(integer_words)
total_number_words = len(integer_words)

#Calculating the freqs for the words
frequencies = {word: count/total_number_words for word, count in word_counts.items()}

#Calculating the discard probability
prob_drop = {word: 1 - np.sqrt(word_threshold/frequencies[word]) for
```

```
word in word_counts}
training_words = [word for word in integer_words if random.random() <
(1 - prob_drop[word])]
```

现在，我们就有了一个筛选后的干净的输入文本。

前面提到过，skip-gram 架构在生成真实数值表示时要考虑目标单词的上下文，因此它在目标单词周围定义了一个大小为 C 的窗口。

比起公平地对待所有上下文单词，我们更倾向于为距离目标单词较远的单词分配较少的权重。例如，如果选择窗口的大小 $C=4$，那么我们将在 $1\sim C$ 中选择一个随机数 L，然后从当前单词的历史版本和未来版本中采样 L 个单词。有关此问题的更多信息，请参考 Mikolov 等人的论文 "Efficient Estimation of Word Representations in Vector Space"。

接下来，定义这个函数。

```
# Defining a function that returns the words around specific index in
a specific window
def get_target(input_words, ind, context_window_size=5):
    #selecting random number to be used for genearting words form history
and feature of the current word
    rnd_num = np.random.randint(1, context_window_size+1)
    start_ind = ind - rnd_num if (ind - rnd_num) > 0 else 0
    stop_ind = ind + rnd_num
    target_words = set(input_words[start_ind:ind] + input_words[ind+1:
stop_ind+1])
    return list(target_words)
```

同时，定义一个生成函数来从训练样本中生成一个随机批次并获得该批次中每个单词的上下文单词。

```
#Defining a function for generating word batches as a tuple (inputs,
targets)
def generate_random_batches(input_words, train_batch_size,
context_window_size=5):
    num_batches = len(input_words)//train_batch_size
    # working on only only full batches
    input_words = input_words[:num_batches*train_batch_size]
    for ind in range(0, len(input_words), train_batch_size):
        input_vals, target = [], []
        input_batch = input_words[ind:ind+train_batch_size]
        #Getting the context for each word
```

```
    for ii in range(len(input_batch)):
        batch_input_vals = input_batch[ii]
        batch_target = get_target(input_batch, ii, context_window_size)
        target.extend(batch_target)
        input_vals.extend([batch_input_vals]*len(batch_target))
    yield input_vals, target
```

11.4.2 构建模型

接下来，我们将用图 11.11 所示架构来构建计算图。

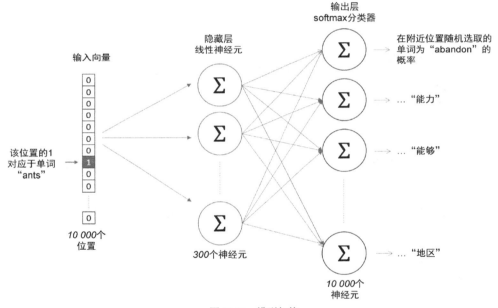

图 11.11 模型架构

所以，正如前面提到的，我们将会使用嵌入层来尝试学习这些单词的一个特殊实值表示。这样，单词将作为一个独热向量传送。我们的想法是训练这个网络来构建权重矩阵。

从创建模型的输入开始。

```
train_graph = tf.Graph()

#defining the inputs placeholders of the model
with train_graph.as_default():
    inputs_values = tf.placeholder(tf.int32, [None], name='inputs_values')
```

```
labels_values = tf.placeholder(tf.int32, [None, None],
name='labels_values')
```

要构建的权重（或者嵌入矩阵）的形状如下。

```
num_words X num_hidden_neurons
```

我们不必自己实现查找函数，因为在 TensorFlow 中已经可以使用 **tf.nn.embedding_lookup()** 了。该函数将使用单词的整数编码并在权重矩阵中定位单词对应的行。

将从一个均匀分布随机初始化权重矩阵。

```
num_vocab = len(integer_to_vocab)

num_embedding =  300
with train_graph.as_default():
    embedding_layer = tf.Variable(tf.random_uniform((num_vocab, num_embedding), -1, 1))
    # Next, we are going to use tf.nn.embedding_lookup function to get the output of the hidden layer
    embed_tensors = tf.nn.embedding_lookup(embedding_layer, inputs_values)
```

一次更新嵌入层的所有嵌入权重是非常低效的。我们可以使用负采样方法，它只使用一小部分不正确的单词来更新正确单词的权重。

同样，我们不需要自己实现这个函数，因为 TensorFlow 里面已经有了 **tf.nn.sampled_softmax_loss**。

```
# Number of negative labels to sample
num_sampled = 100

with train_graph.as_default():
    # create softmax weights and biases
    softmax_weights = tf.Variable(tf.truncated_normal((num_vocab, num_embedding)))
    softmax_biases = tf.Variable(tf.zeros(num_vocab), name="softmax_bias")
    # Calculating the model loss using negative sampling
    model_loss = tf.nn.sampled_softmax_loss(
        weights=softmax_weights,
        biases=softmax_biases,
        labels=labels_values,
        inputs=embed_tensors,
        num_sampled=num_sampled,
        num_classes=num_vocab
```

```
    model_cost = tf.reduce_mean(model_loss)
    model_optimizer = tf.train.AdamOptimizer().minimize(model_cost)
```

为了验证模型，这里将采样一些频繁出现的单词以及一些不常见的单词，并尝试根据 skip-gram 架构习得的表示来输出最接近的单词集。

```
with train_graph.as_default():
    # set of random words for evaluating similarity on
    valid_num_words = 16
    valid_window = 100
    # pick 8 samples from (0,100) and (1000,1100) each ranges. lower id
implies more frequent
    valid_samples = np.array(random.sample(range(valid_window), valid_
num_words//2))
    valid_samples = np.append(valid_samples,
                        random.sample(range(1000,1000+valid_window),
valid_num_words//2))
    valid_dataset_samples = tf.constant(valid_samples, dtype=tf.int32)
    # Calculating the cosine distance
    norm = tf.sqrt(tf.reduce_sum(tf.square(embedding_layer), 1, keep_
dims=True))
    normalized_embed = embedding_layer / norm
    valid_embedding = tf.nn.embedding_lookup(normalized_embed, valid_
dataset_samples)
    cosine_similarity = tf.matmul(valid_embedding,
tf.transpose(normalized_embed))
```

现在，我们有了模型的所有细节，可以准备开始训练了。

11.4.3 训练模型

继续往前进行并开始训练过程。

```
num_epochs = 10
train_batch_size = 1000
contextual_window_size = 10

with train_graph.as_default():
    saver = tf.train.Saver()

with tf.Session(graph=train_graph) as sess:
    iteration_num = 1
    average_loss = 0
```

```python
    #Initializing all the vairables
    sess.run(tf.global_variables_initializer())
    for e in range(1, num_epochs+1):
        #Generating random batch for training
        batches = generate_random_batches(training_words, train_batch_size, contextual_window_size)
        #Iterating through the batch samples
        for input_vals, target in batches:
            #Creating the feed dict
            feed_dict = {inputs_values: input_vals,
                    labels_values: np.array(target)[:, None]}
            train_loss, _ = sess.run([model_cost, model_optimizer], feed_dict=feed_dict)
            #commulating the loss
            average_loss += train_loss
            #Printing out the results after 100 iteration
            if iteration_num % 100 == 0:
                print("Epoch Number {}/{}".format(e, num_epochs),
                    "Iteration Number: {}".format(iteration_num),
                    "Avg. Training loss: {:.4f}".format(average_loss/100))
                average_loss = 0
            if iteration_num % 1000 == 0:
                ## Using cosine similarity to get the nearest words to a word
                similarity = cosine_similarity.eval()
                for i in range(valid_num_words):
                    valid_word = integer_to_vocab[valid_samples[i]]
                    # number of nearest neighbors
                    top_k = 8
                    nearest_words = (-similarity[i, :]).argsort()[1:top_k+1]
                    msg = 'The nearest to %s:' % valid_word
                    for k in range(top_k):
                        similar_word = integer_to_vocab[nearest_words[k]]
                        msg = '%s %s,' % (msg, similar_word)
                    print(msg)
            iteration_num += 1
    save_path = saver.save(sess, "checkpoints/cleaned_wikipedia_version.ckpt")
    embed_mat = sess.run(normalized_embed)
```

在运行了 10 次前面的代码段后，将会得到以下输出。

```
Epoch Number 10/10 Iteration Number: 43100 Avg. Training loss: 5.0380
Epoch Number 10/10 Iteration Number: 43200 Avg. Training loss: 4.9619
Epoch Number 10/10 Iteration Number: 43300 Avg. Training loss: 4.9463
Epoch Number 10/10 Iteration Number: 43400 Avg. Training loss: 4.9728
Epoch Number 10/10 Iteration Number: 43500 Avg. Training loss: 4.9872
Epoch Number 10/10 Iteration Number: 43600 Avg. Training loss: 5.0534
Epoch Number 10/10 Iteration Number: 43700 Avg. Training loss: 4.8261
Epoch Number 10/10 Iteration Number: 43800 Avg. Training loss: 4.8752
Epoch Number 10/10 Iteration Number: 43900 Avg. Training loss: 4.9818
Epoch Number 10/10 Iteration Number: 44000 Avg. Training loss: 4.9251
The nearest to nine: one, seven, zero, two, three, four, eight, five,
The nearest to such: is, as, or, some, have, be, that, physical,
The nearest to who: his, him, he, did, to, had, was, whom,
The nearest to two: zero, one, three, seven, four, five, six, nine,
The nearest to which: as, a, the, in, to, also, for, is,
The nearest to seven: eight, one, three, five, four, six, zero, two,
The nearest to american: actor, nine, singer, actress, musician, comedian,
athlete, songwriter,
The nearest to many: as, other, some, have, also, these, are, or,
The nearest to powers: constitution, constitutional, formally, assembly,
state, legislative, general, government,
The nearest to question: questions, existence, whether, answer, truth,
reality, notion, does,
The nearest to channel: tv, television, broadcasts, broadcasting, radio,
channels, broadcast, stations,
The nearest to recorded: band, rock, studio, songs, album, song,
recording, pop,
The nearest to arts: art, school, alumni, schools, students, university,
renowned, education,
The nearest to orthodox: churches, orthodoxy, church, catholic, catholics,
oriental, christianity, christians,
The nearest to scale: scales, parts, important, note, between, its, see,
measured,
The nearest to mean: is, exactly, defined, denote, hence, are, meaning,
example,

Epoch Number 10/10 Iteration Number: 45100 Avg. Training loss: 4.8466
Epoch Number 10/10 Iteration Number: 45200 Avg. Training loss: 4.8836
Epoch Number 10/10 Iteration Number: 45300 Avg. Training loss: 4.9016
```

```
Epoch Number 10/10 Iteration Number: 45400 Avg. Training loss: 5.0218
Epoch Number 10/10 Iteration Number: 45500 Avg. Training loss: 5.1409
Epoch Number 10/10 Iteration Number: 45600 Avg. Training loss: 4.7864
Epoch Number 10/10 Iteration Number: 45700 Avg. Training loss: 4.9312
Epoch Number 10/10 Iteration Number: 45800 Avg. Training loss: 4.9097
Epoch Number 10/10 Iteration Number: 45900 Avg. Training loss: 4.6924
Epoch Number 10/10 Iteration Number: 46000 Avg. Training loss: 4.8999
The nearest to nine: one, eight, seven, six, four, five, american, two,
The nearest to such: can, example, examples, some, be, which, this, or,
The nearest to who: him, his, himself, he, was, whom, men, said,
The nearest to two: zero, five, three, four, six, one, seven, nine
The nearest to which: to, is, a, the, that, it, and, with,
The nearest to seven: one, six, eight, five, nine, four, three, two,
The nearest to american: musician, actor, actress, nine, singer,
politician, d, one,
The nearest to many: often, as, most, modern, such, and, widely, traditional,
The nearest to powers: constitutional, formally, power, rule, exercised,
parliamentary, constitution, control,
The nearest to question: questions, what, answer, existence, prove,
merely, true, statements,
The nearest to channel: network, channels, broadcasts, stations, cable,
broadcast, broadcasting, radio,
The nearest to recorded: songs, band, song, rock, album, bands, music,
studio,
The nearest to arts: art, school, martial, schools, students, styles,
education, student,
The nearest to orthodox: orthodoxy, churches, church, christianity,
christians, catholics, christian, oriental,
The nearest to scale: scales, can, amounts, depends, tend, are, structural,
for,
The nearest to mean: we, defined, is, exactly, equivalent, denote, number,
above,
Epoch Number 10/10 Iteration Number: 46100 Avg. Training loss: 4.8583
Epoch Number 10/10 Iteration Number: 46200 Avg. Training loss: 4.8887
```

从输出中可以看到，网络在某种程度上学到了输入单词的语义上有用的表示。为了更清楚地了解嵌入矩阵，我们将使用如 t-SNE 的降维法来将实值向量降低到 2 维，然后将它们可视化并在每个点上标记对应的单词。

```
num_visualize_words = 500
tsne_obj = TSNE()
```

```
embedding_tsne =
tsne_obj.fit_transform(embedding_matrix[:num_visualize_words, :])

fig, ax = plt.subplots(figsize=(14, 14))
for ind in range(num_visualize_words):
    plt.scatter(*embedding_tsne[ind, :], color='steelblue')
    plt.annotate(integer_to_vocab[ind], (embedding_tsne[ind, 0],
embedding_tsne[ind, 1]), alpha=0.7)
```

输出结果如图 11.12 所示。

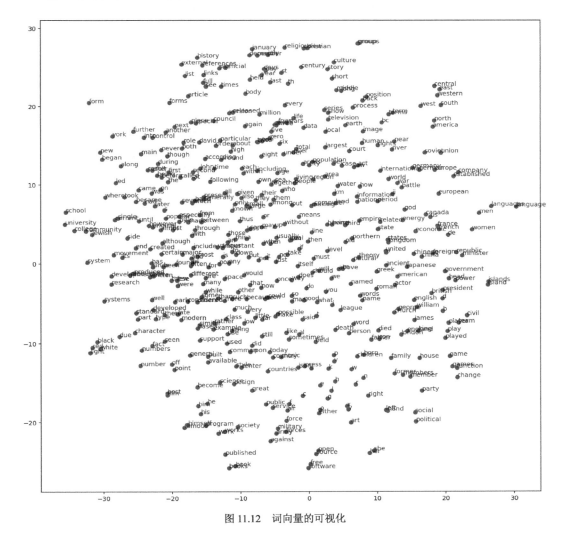

图 11.12 词向量的可视化

11.5 总结

本章讲解了表示学习的概念，以及为什么它对于一些输入不是实值形式的深度学习或机器学习是很有用的。此外，本章还介绍了将单词转换为实值向量的一种技术——Word2Vec，它具有十分有趣的性质。最后，本章使用 skip-gram 架构实现了 Word2Vec 模型。

接下来，我们将在情感分析示例中看到这些习得的表示的实际用法，这需要将输入文本转换为实值向量。

第 12 章
神经网络在情感分析中的应用

本章将讨论自然语言处理中最热门、最流行的应用之一——**情感分析**。现在大多数人通过社交媒体平台表达他们对某事的看法,利用这些文本来跟踪人们对某事物的满意度对于公司甚至政府来说都是至关重要的。

在本章中,我们将使用循环神经网络来构建情感分析解决方案。本章主要讨论以下主题。

- 常用的情感分析模型。
- 情感分析——模型实现。

12.1 常用的情感分析模型

本节将重点介绍可用于情感分析的常用深度学习模型。图 12.1 所示为构建情感分析模型所需的步骤。

所以,首先要处理自然的人类语言。

这里将使用电影评论来构建这种情感分析应用程序。此应用程序的目标是根据输入的原始文本判断它是正面或负面评论。例如,如果原始文本是"This movie is good",那么需要模型把它判断为正面的情感。

情感分析应用程序将带领我们完成在神经网络

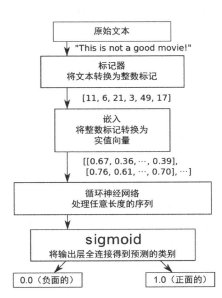

图 12.1 情感分析解决方案或者基于序列的自然语言解决方案的一般流程

中使用自然人类语言所需的许多处理步骤，如嵌入。

在这种情况下，假设有一个原始文本"This is not a good movie!"，我们想要模型把它判断为一种负面或正面的情感。

这类应用有以下困难。

- 序列可能具有**不同的长度**。刚才的例子是一个非常短的句子，接下来我们会看到超过 500 字的文本样本。
- 如果读者只看单个单词（例如，good），那就表明它是正面的情感。然而，因为它之前的单词是 **not**，所以现在它是负面情感。这可能会变得更复杂，我们稍后会看到这方面的一个例子。

正如之前所了解到的，神经网络无法处理原始文本，因此我们需要先将它转换为所谓的**标记**（token）。因为这些标记只包含整数值，所以首先需要遍历整个数据集，并统计每个单词的出现次数。然后，制作一个词汇表，每个单词都从这个词汇表中得到一个索引。例如，单词 **this** 具有一个整数 ID 或标记 **11**，单词 **is** 具有标记 **6**，**not** 具有标记 **21**，依次类推。现在，我们已经将原始文本转换为名为标记的整数列表。

神经网络仍无法对此数据进行操作，因为如果有一个包含 10 000 个单词的词汇表，则标记可以取 0～9999 的值，并且它们可能根本不相关。换句话说，998 号单词可以具有与 999 号单词完全不同的语义。

因此，这里将使用第 11 章中讲到的表示学习或词嵌入的概念。此嵌入层将整数标记转换为实值向量，因此标记 **11** 成为向量[0.67，0.36，…，0.39]，如图 12.1 所示。这同样适用于下一个标记编号 6。

快速回顾一下第 11 章中所研究的内容：图 12.1 中的这个嵌入层学习了标记及其对应的实值向量之间的映射。此外，嵌入层可以学习单词的语义，使得具有相似含义的单词在该嵌入空间中以某种方式接近。

在输入原始文本后，我们得到二维矩阵或张量，现在可以将它输入**循环神经网络**（recurrent neural network，RNN）中了。RNN 可以处理任意长度的序列，然后将该网络的输出传送到具有 sigmoid 激活函数的全连接层或密集层。所以，输出介于 0～1，其中，0 表示负面情感。但是，如果 sigmoid 函数的值既不是 0 也不是 1，该怎么办？这里需要在中间引入一个截止值或阈值，这样如果该值低于 0.5，则相应的输入被视为负面情感，高于此阈值的值被视为正面情感。

12.1.1 RNN——情感分析背景

现在，我们来回顾一下 RNN 的基本概念，并在情感分析应用的背景下讨论它们。正如之前提到的，RNN 的基本构建块是一个循环单元，如图 12.2 所示。

这张图是对循环单位内部情况的抽象。这里有输入，它是一个词，如"**good**"。当然，它必须转换为嵌入向量。这里暂时不考虑这一点。此外，该单元具有一种存储状态，根据该状态和输入的内容，我们将更新该状态并将

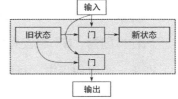

图 12.2 RNN 单元的抽象概念

新数据写入该状态。例如，假设之前输入过"**not**"这个词，当时把它写入存储状态中，这样当在随后的其中一个输入中看到了"**good**"这个单词后，我们就从存储状态中知道刚刚输入了单词"**not**"。现在，我们已经输入了"**good**"这个单词。因此，必须将"**not good**"更新到存储状态中，这表明整个输入文本可能具有负面情感。

从旧状态和输入状态的新内容的映射是通过所谓的门（Gate）完成的，这些实现方式在不同版本的循环单元中是不同的。它基本上是一个具有激活函数的矩阵运算，但正如我们稍后将看到的，反向传播梯度存在一个问题。因此，我们必须以特殊方式设计 RNN，以使梯度不会过度扭曲。

在循环单元中，有一个类似的门用于计算输出，并且之后循环单元的输出取决于状态的当前内容和我们看到的输入。因此，这里可以尝试做的是展开使用循环单元进行的处理，如图 12.3 所示。

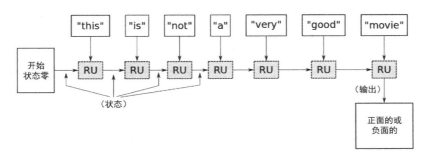

图 12.3 循环神经网络的展开版本

现在，这里只有一个循环单元，但流程图显示了在不同时间步发生的情况。

在时间步 1 中，将单词"**this**"输入循环单元，循环单元的内部存储状态被初始化

为零。对新的数据序列的处理都是由 TensorFlow 完成的。因此，我们可以看到单词"this"和循环单元的状态（即 0）。这里使用内部门更新存储状态，以便"this"在我们输入第二个单词"is"时使用。现在，存储状态中有了一些内容。因为"this"这个单词没有很多实际含义，所以存储状态可能仍然近似为 0。

同样，因为"is"也没有太多含义，所以存储状态依然近似为 0。

在下一个时间步中，我们可以看到单词"not"，这会影响到我们对于整个输入文本情感的最终预测。同样，这也是需要存储在记忆中的内容，以避免循环单元内的门所处理的状态只包含接近零的值。因为现在要存储我们刚看到的单词"not"，所以此时状态会保存一些非零值。

接下来看下一个时间步。在得到单词"a"之后，因为它也没有包含太多信息，所以它可能被忽略了。新的存储状态只是复制了一遍。

现在，我们得到了"very"这个词，这表明存在的情感都可能是一种强烈的情感。到现在为止，循环单元已经以某种方式将单词"not"和"very"存储在其存储状态中了。

在下一个时间步中，我们可以看到单词"good"，因此现在神经网络已经处理了短语"not very good"，并且可能会得出这样的结论"这可能是一个负面情感！"。于是这种想法也会存储在内部状态中。

然后，在最后的时间步中，我们可以看到单词"movie"。因为这个单词相关性不大，所以它可能被忽略。

接下来，使用循环单元内的另一个门来输出内部存储状态的内容，然后用 sigmoid 函数处理它（这里没有显示）。最终我们会得到一个介于 0~1 的输出值。

这里的想法是，希望通过互联网电影数据库对成千上万的电影评论样本进行网络训练。对于每个输入文本，我们给出正面或负面的真实情感值。然后，我们希望 TensorFlow 找出循环单元内的门应该是什么，以便它们准确地将此输入文本映射到正确的情感上，如图 12.4 所示。

我们在此实现中使用的 RNN 的架构是 3 层的 RNN 类型架构。在第一层中，就像刚才解释的，不同点是这里我们需要在每一个时间步从循环单元中输出值。然后，我们收集了一个新的数据序列，它是第一个循环层的输出。接下来，可以将它输入第二个循环层中，因为循环单元需要输入数据序列（我们从第一层获得的输出和要输入第二个循环层的输入是一些浮点值，其实这里我们并没有真正理解它的含义）。这个新的数据序列在

RNN 中有意义，但它不是人类会理解的东西。然后，在第二个循环层中进行类似的处理。

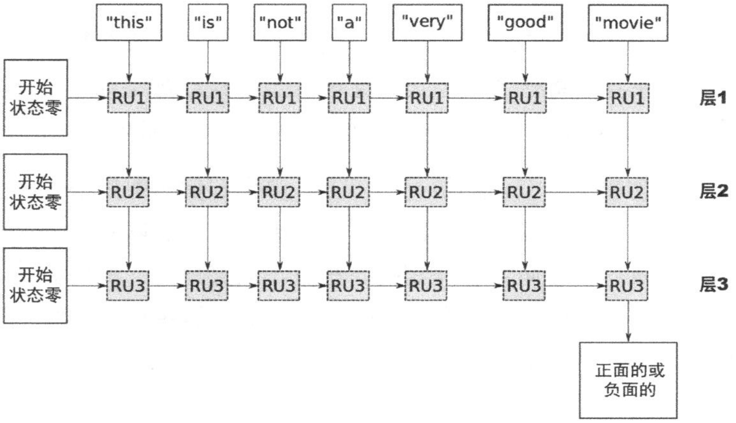

图 12.4　本章所使用的模型结构之一

我们将此循环单元的内部存储状态初始化为 0，然后从第一个循环层获取输出并将它作为输入。这里我们用循环单元内的门处理它，并更新状态，接着获取第一层的循环单元在输入单词"is"后的第二次输出，并将它作为输入继续更新内部存储状态。继续这样做，直到已经处理完整个序列，然后收集第二个循环层的所有输出。这里我们将第二个循环层的输出当作第三个循环层中的输入，在第三层进行类似的处理。但是这里我们只想要最后一个时间步的输出，这是迄今为止所有信息的一种摘要，接下来，将它输出到未在此处显示的全连接层中。最后，我们使用 sigmoid 激活函数，得到了一个介于 0～1 的值，0 代表负面情感，1 代表正面情感。

12.1.2　梯度爆炸与梯度消失——回顾

正如第 11 章提到的，现在存在一种名为梯度**爆炸**和梯度**消失**的现象，这在 RNN 中非常重要。现在回过头来看图 12.1，该流程图解释了这种现象。

想象一下，某数据集中有一段包含 500 个单词的文本，我们将用它们来实现情感分析分类器。在每个时间步中，我们以递归方式在循环单元中应用内部门。因此，如果有 500 个单词，我们将会应用这些门 500 次以更新循环单元的内部存储状态。

我们已经知道，训练神经网络的方式是使用名为梯度反向传播的算法，这里可以通过神经网络的输出和此样本的正确输出计算出损失函数。我们希望最小化该损失函数，以便神经网络的实际输出对应于该特定输入文本的期望输出。因此，我们需要针对这些循环单元内的权重计算此损失函数的梯度，这些权重还可用于更新内部状态并输出最终值的门。

现在，门可能已经运算了 500 次，如果它包含一次乘法，那么基本上得到的是一个指数函数。因此，如果我们将一个值与其自身相乘 500 次，如果该值小于 1，那么它将很快"消失"。同样，如果大于 1 的值与其自身相乘 500 次，那么它会"爆炸"。

唯一能够在 500 次乘法中存活的值是 0 和 1。由于它们将保持不变，因此循环单元实际上比我们在此处看到的要复杂得多。理想的情况是：可以以某种方式映射内部存储状态和输入，从而更新内部存储状态并输出一些值。但实际上，我们需要非常小心地通过这些门向后传播梯度，这样就不会在很多时间步中进行这种指数乘法了。要了解更多信息，可以查看有关循环单位数学定义的一些教程。

12.2 情感分析——模型实现

我们已经详细了解了如何实现 RNN 中变体 LSTM 的堆叠版本。为了让事情更有趣，本节将使用名为 Keras 的更高级的 API。

12.2.1 Keras

"Keras 是一个高级神经网络 API，用 Python 编写，能够在 TensorFlow、CNTK 或 Theano 上运行。它的开发重点是实现快速实验。能够以最短的时间实现一个想法是做好研究的关键。"

——Keras 网站

因此，Keras 只是 TensorFlow 和其他深度学习框架的一个包装。它非常适合于原型设计和快速构建，但是它减少了对代码的控制。我们有机会在 Keras 中实现这种情感分析模型，从而加深对 TensorFlow 和 Keras 的理解。我们可以使用 Keras 进行快速原型设计，使用 TensorFlow 实现生产系统。

更有趣的消息是，我们无须切换到完全不同的环境。现在可以在 TensorFlow 中将 Keras 作为模块访问并导入包。

```
from tensorflow.python.keras.models
import Sequential
from tensorflow.python.keras.layers
```

```
import Dense, GRU, Embedding
from tensorflow.python.keras.optimizers
import Adam
from tensorflow.python.keras.preprocessing.text
import Tokenizer
from tensorflow.python.keras.preprocessing.sequence
import pad_sequences
```

继续使用现在 TensorFlow 中这个更抽象的 Keras 模块,它将帮助我们快速构建深度学习解决方案。我们将用几行代码完成完整的深度学习解决方案。

12.2.2 数据分析与预处理

下面来看如何实现数据加载。Keras 实际上有一个功能可以用来从 IMDb 中加载这个情感数据集,但问题是该函数已经将所有单词映射到整数标记。这是在神经网络中使用自然人类语言的一个重要部分,接下来将介绍如何操作。

此外,如果希望使用此代码对其他语言提供的任何数据进行情感分析,则需要自己执行此操作。我们很快就实现了一些用于下载此数据集的函数。

现在需要导入一堆必需的包。

```
%matplotlib inline
import matplotlib.pyplot as plt
import tensorflow as tf
import numpy as np
from scipy.spatial.distance import cdist
from tensorflow.python.keras.models import Sequential
from tensorflow.python.keras.layers import Dense, GRU, Embedding
from tensorflow.python.keras.optimizers import Adam
from tensorflow.python.keras.preprocessing.text import Tokenizer
from tensorflow.python.keras.preprocessing.sequence import pad_sequences
```

接下来加载数据集。

```
import imdb
imdb.maybe_download_and_extract()
```

输出如下。

```
- Download progress: 100.0%
Download finished. Extracting files.
```

```
Done.

input_text_train, target_train = imdb.load_data(train=True)
input_text_test, target_test = imdb.load_data(train=False)

print("Size of the trainig set: ", len(input_text_train))
print("Size of the testing set:  ", len(input_text_test))
```

输出如下。

```
Size of the trainig set: 25000
Size of the testing set: 25000
```

现在可以看到训练集和测试集各有 25000 个文本。

现在来看训练集中的一个例子。

```
#combine dataset
text_data = input_text_train + input_text_test
input_text_train[1]
```

输出如下。

'This is a really heart-warming family movie. It has absolutely brilliant animal training and "acting" (if you can call it like that) as well (just think about the dog in "How the Grinch stole Christmas"... it was plain bad training). The Paulie story is extremely well done, well reproduced and in general the characters are really elaborated too. Not more to say except that this is a GREAT MOVIE!

My ratings: story 8.5/10, acting 7.5/10, animals+fx 8.5/10, cinematography 8/10.

My overall rating: 8/10 - BIG FAMILY MOVIE AND VERY WORTH WATCHING!'

```
target_train[1]
```

输出如下。

1.0

这是一个相当短的示例，情感值是 1.0，这意味着它是一个正面的情感，因此这是对电影的正面评论。

现在，我们开始使用标记器，这是处理这些原始数据的第一步，因为神经网络无法处理文本数据。Keras 实现了所谓的标记器，用于构建词汇表，从而完成从单词到整数的映射。

此外,假设这里最多需要 10 000 个单词,那么它将只使用数据集中 10 000 个最常用的单词。

```
num_top_words = 10000
tokenizer_obj = Tokenizer(num_words=num_top_words)
```

现在,从数据集中获取所有文本,并在文本中调用 fit 函数。

```
tokenizer_obj.fit_on_texts(text_data)
```

标记器大约需要准备 10s,然后它将构建词汇表,具体如下。

```
tokenizer_obj.word_index
```

输出如下。

```
{'britains': 33206,
 'labcoats': 121364,
 'steeled': 102939,
 'geddon': 67551,
 "rossilini's": 91757,
 'recreational': 27654,
 'suffices': 43205,
 'hallelujah': 30337,
 'mallika': 30343,
 'kilogram': 122493,
 'elphic': 104809,
 'feebly': 32818,
 'unskillful': 91728,
 "'mistress'": 122218,
 "yesterday's": 25908,
 'busco': 85664,
 'goobacks': 85670,
 'mcfeast': 71175,
 'tamsin': 77763,
 "petron's": 72628,
 "'lion": 87485,
 'sams': 58341,
 'unbidden': 60042,
 "principal's": 44902,
```

```
'minutiae': 31453,
'smelled': 35009,
'history\x97but': 75538,
'vehemently': 28626,
'leering': 14905,
'kynay': 107654,
'intendend': 101260,
'chomping': 21885,
'nietsze': 76308,
'browned': 83646,
'grosse': 17645,
"''gaslight''": 74713,
'forseeing': 103637,
'asteroids': 30997,
'peevish': 49633,
"attic'": 120936,
'genres': 4026,
'breckinridge': 17499,
'wrist': 13996,
"sopranos'": 50345,
'embarasing': 92679,
"wednesday's": 118413,
'cervi': 39092,
'felicity': 21570,
"''horror''": 56254,
'alarms': 17764,
"'ol": 29410,
'leper': 27793,
'once\x85': 100641,
'iverson': 66834,
'triply': 117589,
'industries': 19176,
'brite': 16733,
'amateur': 2459,
"libby's": 46942,
'eeeeevil': 120413,
'jbc33': 51111,
'wyoming': 12030,
'waned': 30059,
'uchida': 63203,
'uttter': 93299,
'irector': 123847,
```

```
'outriders': 95156,
'perd': 118465,
.
.
.}
```

这样,每一个单词都有一个对应的整数,例如,单词"the"对应数字 1。

tokenizer_obj.word_index['the']

输出如下。

1

同样,"and"对应的数字是 2。

tokenizer_obj.word_index['and']

输出如下。

2

单词"a"对应的数字是 3。

tokenizer_obj.word_index['a']

输出如下。

3

依次类推,可以看到单词"movie"对应的是数字 17。

tokenizer_obj.word_index['movie']

输出如下。

17

单词"film"对应的数字是 19。

tokenizer_obj.word_index['film']

输出如下。

19

所有这些意味着"the"是数据集中出现次数最多的单词,"and"是数据集中出现次数第二多的单词。因此,每当我们想要将单词映射到整数标记时,总会得到这些数字。

下面尝试以数字 743 作为例子,可以发现它对应的单词是"romantic"。

```
tokenizer_obj.word_index['romantic']
```

输出如下。

743

因此,每当我们在输入文本中看到单词"romantic"时,就将它映射到标记整数 743。下面再次使用标记器将训练集里文本中的所有单词转换为整数标记。

```
input_text_train[1]
```

输出如下。

```
'This is a really heart-warming family movie. It has absolutely brilliant
animal training and "acting" (if you can call it like that) as well
(just think about the dog in "How the Grinch stole Christmas"... it was
plain bad training). The Paulie story is extremely well done, well
reproduced and in general the characters are really elaborated too. Not
more to say except that this is a GREAT MOVIE!<br /><br />My ratings:
story 8.5/10, acting 7.5/10, animals+fx 8.5/10, cinematography 8/10.<br />
<br />My overall rating: 8/10 - BIG FAMILY MOVIE AND VERY WORTH WATCHING!
```

当将该文本转换为整数标记时,它将变为整数数组。

```
np.array(input_train_tokens[1])
```

输出如下。

```
array([  11,    6,    3,   62,  488, 4679,  236,   17,    9,   45,  419,
        513, 1717, 2425,    2,  113,   43,   22,   67,  654,    9,   37,
         12,   14,   69,   39,  101,   42,    1,  826,    8,   85,    1,
       6418, 3492, 1156,    9,   13, 1042,   74, 2425,    1, 6419,   64,
          6,  568,   69,  221,   69,    2,    8,  825,    1,  102,   23,
         62,   96,   21,   51,    5,  131,  556,   12,   11,    6,    3,
         78,   17,    7,    7,   56, 2818,   64,  723,  447,  156,  113,
```

```
                702, 447, 156, 1598, 3611, 723, 447, 156, 633, 723, 156,
            7, 7, 56, 437, 670, 723, 156, 191, 236, 17, 2,
            52, 278, 147])
```

单词"this"变成数字 11,"is"变成数字 59。

我们还需要转换文本的其余部分。

```
input_test_tokens = tokenizer_obj.texts_to_sequences(input_text_test)
```

现在,还有另一个问题,因为即使循环单元可以使用任意长度的序列,标记序列的长度也取决于原始文本的长度。然而,TensorFlow 的工作方式是一批中的所有数据都需要具有相同的长度。

因此,要么确保整个数据集中的所有序列具有相同的长度,要么编写自定义数据生成器以确保单个批次中的序列具有相同的长度。现在,确保数据集中的所有序列具有相同的长度要简单得多,但还是存在一些异常值。例如,有些句子的长度超过 2200 个单词。如果所有短句子的长度都超过 2200 个单词,那么将非常消耗内存。所以这里要做的就是妥协。首先,需要统计每个输入序列中的所有单词或标记的数量。可以看到,序列中的平均单词数约为 221。

```
total_num_tokens = [len(tokens) for tokens in input_train_tokens + 
input_test_tokens]
total_num_tokens = np.array(total_num_tokens)

#Get the average number of tokens
np.mean(total_num_tokens)
```

输出如下。

221.27716

最大单词数超过 2200。

```
np.max(total_num_tokens)
```

输出如下。

2208

现在,平均值和最大值之间存在巨大差异,如果只填充数据集中的所有句子,那么会浪费大量内存,这样它们都会拥有 2208 个标记。如果某一数据集包含数百万个文本序列,那么这将成为一个问题。

所以我们要做的是妥协:填充所有序列,并截断那些太长的序列,使它们有 544 个单词。计算的方式为采用数据集中所有序列的平均单词数,并添加两个标准偏差。

```
max_num_tokens = np.mean(total_num_tokens) + 2 * np.std(total_num_tokens)
max_num_tokens = int(max_num_tokens)
max_num_tokens
```

输出如下。

```
544
```

从中得到了什么?我们覆盖了数据集中大约 95%的文本,因此只有大约 5%的文本中的单词数多于 544。

```
np.sum(total_num_tokens < max_num_tokens) / len(total_num_tokens)
```

输出如下。

```
0.94532
```

现在,在 Keras 中调用这些函数。要么填充长度太短的序列(只会添加零),要么截断太长的序列(如果文本太长,去掉一些单词)。

现在,还有一件重要的事情:我们可以在开头或结尾执行此填充和截断。想象一下,有一个整数标记序列,我们需要填充它,因为它太短了。要么在开头填充这些零,以便在结尾处有实际的整数标记。要么以相反的方式进行:在开头拥有所有这些数据,然后在结尾添加所有零。但是如果回过头来看看前面的 RNN 流程图,请记住它一次一步地处理序列,所以如果一开始处理零,它可能不会有任何意义,内部状态可能仍为零。因此,每当它看到特定单词的整数标记时,它就会知道现在开始处理数据了。

然而,如果所有的零都在结尾,那么就会开始处理所有数据,然后在循环单位中有一些内部状态。现在,我们看到了很多零,因此这实际上可能会破坏刚刚计算出来的内部状态。这就是最好选择在开头填充零的原因。

另一个问题是在截断文本时,如果文本很长,我们会截断它以使其长度为 544 个单词,或者为其他设定的单词数。现在,想象一下当看到截断后的句子是"**this very good movie**"或者是"**this is not**"时,我们会如何判断。当然,我们知道只对非常长的序列执行截断操作,但这样做可能会丢失正确分类此文本的必要信息。因此,当我们截断输入文本时,其实正在做出妥协。更好的方法是创建一个批次并在该批次中填充文本。因此,当看到非常长的序列时,我们将其他序列填充为具有相同长度的序列。但是不需要将所有这些数据存储在内存中,因为大部分内存都浪费了。

现在返回并转换整个数据集,以便进行截断和填充。最终,它会变成一个很大的数据矩阵。

```
seq_pad = 'pre'

input_train_pad = pad_sequences(input_train_tokens, maxlen=max_num_tokens,
padding=seq_pad, truncating=seq_pad)

input_test_pad = pad_sequences(input_test_tokens, maxlen=max_num_tokens,
padding=seq_pad, truncating=seq_pad)
```

下面检查一下矩阵的形状。

```
input_train_pad.shape
```

输出如下。

```
(25000, 544)
```

```
input_test_pad.shape
```

输出如下。

```
(25000, 544)
```

下面查看具体样本标记在填充前后的样子。

```
np.array(input_train_tokens[1])
```

输出如下。

```
array([ 11,    6,    3,   62,  488, 4679,  236,   17,    9,   45,  419,
        513, 1717, 2425,    2,  113,   43,   22,   67,  654,    9,   37,
         12,   14,   69,   39,  101,   42,    1,  826,    8,   85,    1,
       6418, 3492, 1156,    9,   13, 1042,   74, 2425,    1, 6419,   64,
          6,  568,   69,  221,   69,    2,    8,  825,    1,  102,   23,
         62,   96,   21,   51,    5,  131,  556,   12,   11,    6,    3,
```

```
        78,   17,    7,    7,   56, 2818,   64,  723,  447,  156,  113,
       702,  447,  156, 1598, 3611,  723,  447,  156,  633,  723,  156,
         7,    7,   56,  437,  670,  723,  156,  191,  236,   17,    2,
        52,  278,  147])
```

下面查看填充之后的样本。

input_train_pad[1]

输出如下。

```
array([ 0,  0,  0,  0,  0,  0,  0,  0,  0,  0,  0,
        0,  0,  0,  0,  0,  0,  0,  0,  0,  0,  0,
        0,  0,  0,  0,  0,  0,  0,  0,  0,  0,  0,
        0,  0,  0,  0,  0,  0,  0,  0,  0,  0,  0,
        0,  0,  0,  0,  0,  0,  0,  0,  0,  0,  0,
        0,  0,  0,  0,  0,  0,  0,  0,  0,  0,  0,
        0,  0,  0,  0,  0,  0,  0,  0,  0,  0,  0,
        0,  0,  0,  0,  0,  0,  0,  0,  0,  0,  0,
        0,  0,  0,  0,  0,  0,  0,  0,  0,  0,  0,
        0,  0,  0,  0,  0,  0,  0,  0,  0,  0,  0,
        0,  0,  0,  0,  0,  0,  0,  0,  0,  0,  0,
        0,  0,  0,  0,  0,  0,  0,  0,  0,  0,  0,
        0,  0,  0,  0,  0,  0,  0,  0,  0,  0,  0,
        0,  0,  0,  0,  0,  0,  0,  0,  0,  0,  0,
        0,  0,  0,  0,  0,  0,  0,  0,  0,  0,  0,
        0,  0,  0,  0,  0,  0,  0,  0,  0,  0,  0,
        0,  0,  0,  0,  0,  0,  0,  0,  0,  0,  0,
        0,  0,  0,  0,  0,  0,  0,  0,  0,  0,  0,
        0,  0,  0,  0,  0,  0,  0,  0,  0,  0,  0,
        0,  0,  0,  0,  0,  0,  0,  0,  0,  0,  0,
        0,  0,  0,  0,  0,  0,  0,  0,  0,  0,  0,
        0,  0,  0,  0,  0,  0,  0,  0,  0,  0,  0,
        0,  0,  0,  0,  0,  0,  0,  0,  0,  0,  0,
        0,  0,  0,  0,  0,  0,  0,  0,  0,  0,  0,
        0,  0,  0,  0,  0,  0,  0,  0,  0,  0,  0,
```

```
         0,    0,    0,    0,    0,    0,    0,    0,    0,    0,    0,
         0,    0,    0,    0,    0,    0,    0,    0,    0,    0,    0,
         0,    0,    0,    0,    0,    0,    0,    0,    0,    0,    0,
         0,    0,    0,    0,    0,    0,    0,    0,    0,    0,    0,
         0,    0,    0,    0,    0,    0,    0,    0,    0,    0,    0,
         0,    0,    0,    0,    0,    0,    0,    0,    0,    0,    0,
         0,    0,    0,    0,    0,    0,    0,    0,    0,    0,    0,
         0,    0,    0,    0,    0,    0,    0,    0,    0,    0,    0,
         0,    0,   11,    6,    3,   62,  488, 4679,  236,   17,    9,
        45,  419,  513, 1717, 2425,    2,  113,   43,   22,   67,  654,
         9,   37,   12,   14,   69,   39,  101,   42,    1,  826,    8,
        85,    1, 6418, 3492, 1156,    9,   13, 1042,   74, 2425,    1,
      6419,   64,    6,  568,   69,  221,   69,    2,    8,  825,    1,
       102,   23,   62,   96,   21,   51,    5,  131,  556,   12,   11,
         6,    3,   78,   17,    7,    7,   56, 2818,   64,  723,  447,
       156,  113,  702,  447,  156, 1598, 3611,  723,  447,  156,  633,
       723,  156,    7,    7,   56,  437,  670,  723,  156,  191,  236,
        17,    2,   52,  278,  147], dtype=int32)
```

此外,我们还需要一个反向映射的功能,以便能从整数标记映射回文本单词。这是一个非常简单的辅助函数,现在实现它。

```
index = tokenizer_obj.word_index
index_inverse_map = dict(zip(index.values(), index.keys()))

def convert_tokens_to_string(input_tokens):

 # Convert the tokens back to words
 input_words = [index_inverse_map[token] for token in input_tokens if token != 0]

 # join them all words.
 combined_text = " ".join(input_words)

return combined_text
```

例如,数据集中的原始文本如下。

```
input_text_train[1]
```

输出如下。

```
input_text_train[1]
```

'This is a really heart-warming family movie. It has absolutely brilliant animal training and "acting" (if you can call it like that) as well (just think about the dog in "How the Grinch stole Christmas"... it was plain bad training). The Paulie story is extremely well done, well reproduced and in general the characters are really elaborated too. Not more to say except that this is a GREAT MOVIE!

My ratings: story 8.5/10, acting 7.5/10, animals+fx 8.5/10, cinematography 8/10.

My overall rating: 8/10 - BIG FAMILY MOVIE AND VERY WORTH WATCHING!'

如果使用辅助函数将标记转换回文本单词，将得到以下文本。

```
convert_tokens_to_string(input_train_tokens[1])
```

'this is a really heart warming family movie it has absolutely brilliant animal training and acting if you can call it like that as well just think about the dog in how the grinch stole christmas it was plain bad training the paulie story is extremely well done well and in general the characters are really too not more to say except that this is a great movie br br my ratings story 8 5 10 acting 7 5 10 animals fx 8 5 10 cinematography 8 10 br br my overall rating 8 10 big family movie and very worth watching'

除了标点符号和其他符号外，以上文本与原文本基本相同。

12.2.3 构建模型

现在需要创建 RNN，我们将在 Keras 中执行此操作，过程非常简单。我们用所谓的 sequential 模型来实现。

这种架构的第一层是**嵌入**（embedding）层。回顾图 12.1 所示的流程，我们刚刚做的是将原始输入文本转换为整数标记。但我们仍然无法将它输入 RNN 中，因此必须将它转换为嵌入向量，嵌入向量的值介于 $-1\sim1$。在某种程度上向量的值可以超出该范围，但通常介于 $-1\sim1$，这是可以在神经网络中使用的数据。

神奇之处在于，这个嵌入层与 RNN 同时训练并且 RNN 无法识别原始单词。它只能识别整数标记，但需要学会认识单词之间的使用规则。因此，嵌入层推断某些词或某些整数标记可能具有相似的含义，然后在嵌入向量中对这些看起来相似的单词进行编码。

现在，我们需要确定的是每个向量的长度。例如，将标记"11"转换为实值向量。在这个例子中，将使用长度为 8 的嵌入向量，实际上这个值非常短（通常，它介于 100～300）。现在尝试在嵌入向量中更改此属性，然后重新运行代码以查看结果。

因此，我们将嵌入向量的长度设置为 8，然后使用 Keras 将此嵌入层添加到 RNN 中。它必须是网络的第一层。

```
embedding_layer_size = 8

rnn_type_model.add(Embedding(input_dim=num_top_words,
                    output_dim=embedding_layer_size,
                    input_length=max_num_tokens,
                    name='embedding_layer'))
```

然后，可以添加第一个循环层了，这里将使用**门控循环单元**（Gated Recurrent Unit，GRU）。通常，我们会看到很多人使用 LSTM 网络，但有一些人似乎认为 GRU 更好，因为 LSTM 网络内部有些门是多余的。事实上，使用更少的门，也可以使代码变得简单。我们可以为 LSTM 网络增加一千多个门，但这并不意味着它会变得更好。

现在开始定义 GRU 架构，假设输出维度为 16，返回的序列模型如下。

```
rnn_type_model.add(GRU(units=16, return_sequences=True))
```

看一下图 12.4 所示的流程图，我们还想要添加第二个循环层。

```
rnn_type_model.add(GRU(units=8, return_sequences=True))
```

然后，添加第三个（即最后一个）循环层，这一层不会输出序列，因为它后面还有一个密集层。它只给出 GRU 的最终输出，而不是整个输出序列。

```
rnn_type_model.add(GRU(units=4))
```

然后，把这里的输出馈送到全连接层或密集层，该层应该为每个输入序列输出一个值。这是使用 sigmoid 激活函数处理过的，因此它输出 0～1 的一个值。

```
rnn_type_model.add(Dense(1, activation='sigmoid'))
```

这里我们使用具有特定学习速率的 Adam 优化器，损失函数是 RNN 的输出和训练集的实际类别值之间的二元交叉熵（binary cross-entropy），这里实际类别值为 0 或 1。

```
model_optimizer = Adam(lr=1e-3)

rnn_type_model.compile(loss='binary_crossentropy',
          optimizer=model_optimizer,
          metrics=['accuracy'])
```

下面输出模型的实际结构。

```
rnn_type_model.summary()
```

```
Layer (type)  Output Shape  Param #
=================================================================
embedding_layer (Embedding) (None, 544, 8) 80000
_____
gru_1 (GRU)  (None, None, 16) 1200
_____
gru_2 (GRU)  (None, None, 8) 600
_____
gru_3 (GRU)  (None, 4) 156
_____
dense_1 (Dense) (None, 1) 5
=================================================================
Total params: 81,961
Trainable params: 81,961
Non-trainable params: 0
```

可以看到，该模型有一个嵌入层、3 个循环层和一个密集层。注意，它们的参数并不多。

12.2.4 模型训练和结果分析

下面开始训练，这很容易。

```
rnn_type_model.fit(input_train_pad, target_train,
          validation_split=0.05, epochs=3, batch_size=64)
```

输出如下。

```
Train on 23750 samples, validate on 1250 samples
Epoch 1/3
23750/23750 [==============================]23750/23750
```

```
[==============================] - 176s 7ms/step - loss: 0.6698 - acc:
0.5758 - val_loss: 0.5039 - val_acc: 0.7784

Epoch 2/3
23750/23750 [==============================]23750/23750
[==============================] - 175s 7ms/step - loss: 0.4631 - acc:
0.7834 - val_loss: 0.2571 - val_acc: 0.8960

Epoch 3/3
23750/23750 [==============================]23750/23750
[==============================] - 174s 7ms/step - loss: 0.3256 - acc:
0.8673 - val_loss: 0.3266 - val_acc: 0.8600
```

接下来，使用训练好的模型测试测试集。

```
model_result = rnn_type_model.evaluate(input_test_pad, target_test)
```

输出如下。

```
25000/25000 [==============================]25000/25000
[==============================] - 60s 2ms/step

print("Accuracy: {0:.2%}".format(model_result[1]))
Output:
Accuracy: 85.26%
```

现在，查看一些错误分类文本的例子。

首先，计算测试集中前 1000 个序列的预测类别，然后获取真实的类别值。对它们进行比较，并获得一个不完全匹配的索引列表。

```
target_predicted = rnn_type_model.predict(x=input_test_pad[0:1000])
target_predicted = target_predicted.T[0]
```

使用阈值来定义高于 0.5 的所有值都将被视为正面情感，而其他值将被视为负面情感。

```
class_predicted = np.array([1.0 if prob>0.5 else 0.0 for prob in target_predicted])
```

现在，获取这 1000 个序列的实际类别。

```
class_actual = np.array(target_test[0:1000])
```

从输出中得到不正确的样本。

```
incorrect_samples = np.where(class_predicted != class_actual)
incorrect_samples = incorrect_samples[0]
len(incorrect_samples)
```

输出如下。

122

可以看到，这些文本中有 122 个文本被错误分类。也就是说，在预测的 1000 个文本中，有 12.2% 的文本被错误分类[①]。下面来看看第一个被错误分类的文本。

```
index = incorrect_samples[0]
index
```

输出如下。

9

```
incorrectly_predicted_text = input_text_test[index]
incorrectly_predicted_text
```

输出如下。

```
'I am not a big music video fan. I think music videos take away personal
feelings about a particular song.. Any song. In other words, creative
thinking goes out the window. Likewise, Personal feelings aside about
MJ, toss aside. This was the best music video of alltime. Simply wonderful.
It was a movie. Yes folks it was. Brilliant! You had awesome acting,
awesome choreography, and awesome singing. This was spectacular. Simply
a plot line of a beautiful young lady dating a man, but was he a man
or something sinister. Vincent Price did his thing adding to the song
and video. MJ was MJ, enough said about that. This song was to video,
what Jaguars are for cars. Top of the line, PERFECTO. What was even
better about this was, that we got the real MJ without the thousand
facelifts. Though ironically enough, there was more than enough makeup
and costumes to go around. Folks go to Youtube. Take 14 mins. out of
your life and see for yourself what a wonderful work of art this particular
video really is.'
```

下面查看这个样本对应的模型输出值以及真实值。

---

① 原文此处为"12.1% 的文本被错误分类"，应为作者笔误。——译者注

```
target_predicted[index]
```

输出如下。

```
0.1529513
```

```
class_actual[index]
```

输出如下。

```
1.0
```

下面根据一组新的数据样本来测试训练的模型并查看结果。

```
test_sample_1 = "This movie is fantastic! I really like it because it
is so good!"
test_sample_2 = "Good movie!"
test_sample_3 = "Maybe I like this movie."
test_sample_4 = "Meh ..."
test_sample_5 = "If I were a drunk teenager then this movie might be good."
test_sample_6 = "Bad movie!"
test_sample_7 = "Not a good movie!"
test_sample_8 = "This movie really sucks! Can I get my money back please?"
test_samples = [test_sample_1, test_sample_2, test_sample_3, test_sample_4,
test_sample_5, test_sample_6, test_sample_7, test_sample_8]
```

然后将它们转化为整数标记。

```
test_samples_tokens = tokenizer_obj.texts_to_sequences(test_samples)
```

下一步进行填充操作。

```
test_samples_tokens_pad = pad_sequences(test_samples_tokens,
maxlen=max_num_tokens,
                        padding=seq_pad, truncating=seq_pad)
test_samples_tokens_pad.shape
```

输出如下。

```
(8, 544)
```

最后,将样本输入模型中。

```
rnn_type_model.predict(test_samples_tokens_pad)
```

输出如下。

```
array([[0.9496784 ],
 [0.9552593 ],
 [0.9115685 ],
 [0.9464672 ],
 [0.87672734],
 [0.81883633],
 [0.33248223],
 [0.15345531 ]], dtype=float32)
```

接近 0 的值意味着负面情感，而接近 1 的值意味着正面情感。最后，注意在每次训练模型时，这些数字都会有所不同。

12.3 总结

本章介绍了一个有趣的应用程序——情感分析。很多公司使用情感分析来跟踪客户对其产品的满意度，甚至政府也使用情感分析来追踪公民对他们未来想要做的事情的满意度。

接下来，我们将重点介绍一些可用于半监督和无监督应用的高级深度学习架构。

第 13 章
自动编码器——特征提取和降噪

自动编码器（autoencoder）网络是目前广泛使用的深度学习架构之一。它主要用于高效解码任务的无监督学习，还可以通过学习特定数据集的编码或表示来降维。在本章中，将通过使用自动编码器，展示如何通过构建具有相同维度但噪声较小的数据集对另一个数据集进行去噪。为了在实践中使用这个概念，这里将从 MNIST 数据集中提取重要特征，并尝试以此来显著增强性能。

本章将介绍以下内容。

- 自动编码器简介。
- 自动编码器的示例。
- 自动编码器架构。
- 压缩 MNIST 数据集。
- 卷积自动编码器。
- 降噪自动编码器。
- 自动编码器的应用。

13.1 自动编码器简介

自动编码器是另一种可用于许多有趣任务的深度学习架构，但它也可以被视为原始前馈神经网络的变体，其中，输出具有与输入相同的维度。如图 13.1 所示，自动编码器的工作方式是将数据样本（$x_1, \cdots, x_6$）提供给网络，它将尝试在 L_2 层中学习此数据的低维表示，我们可以将它称为以低维表示形式对数据集进行编码的方法。然后，网络的第二部分（可称为解码器）负责构造此表示$(\hat{x}_1,\cdots,\hat{x}_6)$的输出。可以将网络从输入数据中学习的中间低维表示视为其压缩版本。

与我们迄今为止看到的所有其他深度学习架构没有太大差别，自动编码器也同样使用反向传播算法。

自动编码器神经网络是一种无监督学习算法，它应用反向传播，将目标值设置为与输入相等的值。

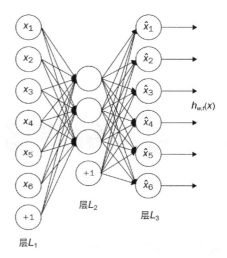

图 13.1　自动编码器的通用结构

13.2　自动编码器的示例

本章将演示使用 MNIST 数据集的自动编码器的多种变体的一些示例。作为具体示例，假设输入 x 是来自 28×28 像素（784 像素）图像的像素值。输入数据样本的数量是 n = 784，L_2 层中有 s_2 = 392 个隐藏单位。由于输出将与输入数据样本具有相同的维度，因此 $y \in R^{784}$。输入层中的神经元数量为 784，其次是中间层 L_2 中的 392 个神经元。现在网络计算出一个低维表示，这是输入的是压缩版本。然后，网络将输入的压缩版本 $a(L_2) \in R^{392}$ 传送到网络的第二部分，这里尽可能用该压缩版本重建输入的 784 个像素。

自动编码器依赖于以下事实：由图像像素表示的输入样本将以某种方式相关联，然后自动编码器将使用该事实来重构它们。自动编码器有点类似于降维法，因为它们也学习了输入数据的低维表示。

总结一下，典型的自动编码器将由如下 3 部分组成。

- 编码器部分，负责将输入压缩到低维空间。
- 中间码，它是编码器的中间结果，可以理解为输入的低维表示。
- 解码器，负责使用中间码重建原始输入。

图 13.2 所示为典型自动编码器的 3 个主要部分。

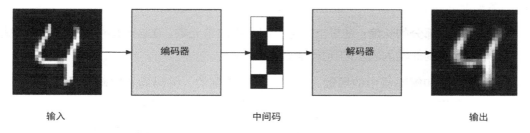

图 13.2　自动编码器运算过程

正如我们所提到的,自动编码器会学习输入的压缩表示,然后将它输入第三部分,第三部分尝试重建输入。重建的输入类似于原始的输入,但它不能保证与原始输入完全相同,因此自动编码器不能用于压缩任务。

13.3　自动编码器架构

如前所述,典型的自动编码器由 3 部分组成,下面更详细地讨论这 3 个部分。为了激励读者,不打算在本章重新发明轮子。如图 13.3 所示,编码器-解码器部分只是一个完全连接的神经网络,中间码部分是另一个神经网络,但它不是全连接网络。中间码部分的维度是可控的,可以将它视为超参数。

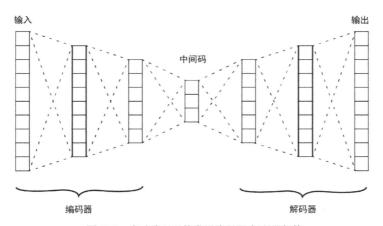

图 13.3　自动编码器的常用编码器-解码器架构

在深入使用自动编码器压缩 MNIST 数据集之前,这里将列出一组可用于微调自动编码器模型的超参数。主要有以下 4 个超参数。

- **中间码部分的长度**：这里指的是中间层中的单元数。在这一层中拥有的单位数越少，得到的输入表示的压缩程度就越高。
- **编码器和解码器中的层数**：正如前文所提到的，编码器和解码器只不过是一个完全连接的神经网络，可以通过添加更多层来增加网络深度。
- **每层的单位数**：可以在每一层中使用不同数量的单位。编码器和解码器的形状与 DeconvNets 非常相似，其中，编码器中的层数越接近中间码部分就越小，而当接近解码器的最后一层时就开始增加。
- **模型损失函数**：可以使用不同的损失函数，如 MSE 或交叉熵（cross-entropy）。

在定义了这些超参数并指定它们的初始值之后，可以使用反向传播算法来训练网络。

13.4　压缩 MNIST 数据集

在本节中，我们将构建一个简单的自动编码器，它可用于压缩 MNIST 数据集。首先，将此数据集的图像提供给编码器部分，编码器部分将尝试为它们学习低维的压缩表示。然后，我们将尝试在解码器部分再次构造输入图像。

13.4.1　MNIST 数据集

我们通过使用 TensorFlow 的辅助函数获取 MNIST 数据集来开始实现自动编码器。

首先，为这个实现导入必要的包。

```
%matplotlib inline

import numpy as np
import tensorflow as tf
import matplotlib.pyplot as plt
from tensorflow.examples.tutorials.mnist import input_data
mnist_dataset = input_data.read_data_sets('MNIST_data', validation_size=0)
```

输出如下。

```
Extracting MNIST_data/train-images-idx3-ubyte.gz
Extracting MNIST_data/train-labels-idx1-ubyte.gz
```

```
Extracting MNIST_data/t10k-images-idx3-ubyte.gz
Extracting MNIST_data/t10k-labels-idx1-ubyte.gz
```

然后，展示数据集中的一些样本图像。

```
# Plotting one image from the training set.
image = mnist_dataset.train.images[2]
plt.imshow(image.reshape((28, 28)), cmap='Greys_r')
```

输出如图 13.4 所示。

```
# Plotting one image from the training set.
image = mnist_dataset.train.images[2]
plt.imshow(image.reshape((28, 28)), cmap='Greys_r')
```

输出如图 13.5 所示。

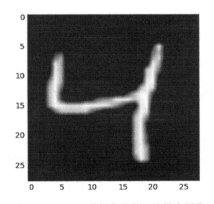 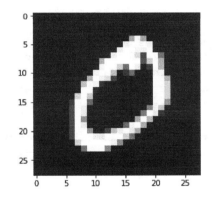

图 13.4　MNIST 数据集中的一些样本图像　　　图 13.5　MNIST 数据集中的一些样本图像

13.4.2　构建模型

为了构建编码器，首先需要弄清楚每幅 MNIST 图像具有多少像素，以便计算出编码器输入层的大小。来自 MNIST 数据集的每幅图像是 28×28 像素，因此将该矩阵转换为一个包含 28×28 = 784 个像素值的向量。这里没有归一化 MNIST 的图像，因为它们已经归一化了。

现在开始构建模型的 3 个部分。在此实现中，我们将使用一个非常简单的体系结构，其中有单个隐藏层，然后是激活 ReLU 函数，如图 13.6 所示。

第 13 章 自动编码器——特征提取和降噪

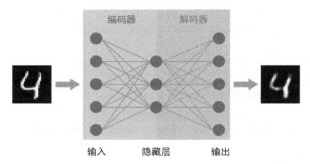

图 13.6 用 MNIST 数据集实现的编码器-解码器架构

下面继续按照前面的解释来实现这个简单的编码器-解码器架构。

```
# The size of the encoding layer or the hidden layer.
encoding_layer_dim = 32

img_size = mnist_dataset.train.images.shape[1]

# defining placeholder variables of the input and target values
inputs_values = tf.placeholder(tf.float32, (None, img_size), name="inputs_values")
targets_values = tf.placeholder(tf.float32, (None, img_size), name="targets_values")

# Defining an encoding layer which takes the input values and incode them.
encoding_layer = tf.layers.dense(inputs_values, encoding_layer_dim, activation=tf.nn.relu)

# Defining the logit layer, which is a fully-connected layer but without
any activation applied to its output
logits_layer = tf.layers.dense(encoding_layer, img_size, activation=None)

# Adding a sigmoid layer after the logit layer
decoding_layer = tf.sigmoid(logits_layer, name = "decoding_layer")

# use the sigmoid cross entropy as a loss function
model_loss = tf.nn.sigmoid_cross_entropy_with_logits(logits=logits_layer, labels=targets_values)

# Averaging the loss values accross the input data
model_cost = tf.reduce_mean(model_loss)
```

```
# Now we have a cost function that we need to optimize using Adam Optimizer
model_optimizier = tf.train.AdamOptimizer().minimize(model_cost)
```

既然图像、像素已经归一化了,现在就可以定义模型了,也可以使用二元交叉熵了。

13.4.3 训练模型

本节将开始进入训练阶段。这里将使用 mnist_dataset 对象的辅助函数,以便从特定大小的数据集中获取一批随机图像,然后在这批图像上运行优化器。

现在通过创建会话变量来开始本节,该变量将负责执行之前定义的计算图。

```
# creating the session
 sess = tf.Session()
```

接下来,进入训练阶段。

```
num_epochs = 20
train_batch_size = 200

sess.run(tf.global_variables_initializer())
for e in range(num_epochs):
    for ii in range(mnist_dataset.train.num_examples//train_batch_size):
        input_batch = mnist_dataset.train.next_batch(train_batch_size)
        feed_dict = {inputs_values: input_batch[0], targets_values: input_batch[0]}
        input_batch_cost, _=sess.run([model_cost,model_optimizier],feed_dict=feed_dict)

        print("Epoch: {}/{}...".format(e+1, num_epochs),
            "Training loss:{:.3f}".format(input_batch_cost))
```

输出如下。

```
.
.
.
Epoch: 20/20... Training loss: 0.091
Epoch: 20/20... Training loss: 0.091
Epoch: 20/20... Training loss: 0.093
Epoch: 20/20... Training loss: 0.093
Epoch: 20/20... Training loss: 0.095
Epoch: 20/20... Training loss: 0.095
Epoch: 20/20... Training loss: 0.089
```

```
Epoch: 20/20... Training loss: 0.095
Epoch: 20/20... Training loss: 0.095
Epoch: 20/20... Training loss: 0.096
Epoch: 20/20... Training loss: 0.094
Epoch: 20/20... Training loss: 0.093
Epoch: 20/20... Training loss: 0.094
Epoch: 20/20... Training loss: 0.093
Epoch: 20/20... Training loss: 0.095
Epoch: 20/20... Training loss: 0.094
Epoch: 20/20... Training loss: 0.096
Epoch: 20/20... Training loss: 0.092
Epoch: 20/20... Training loss: 0.093
Epoch: 20/20... Training loss: 0.091
Epoch: 20/20... Training loss: 0.093
Epoch: 20/20... Training loss: 0.091
Epoch: 20/20... Training loss: 0.095
Epoch: 20/20... Training loss: 0.094
Epoch: 20/20... Training loss: 0.091
Epoch: 20/20... Training loss: 0.096
Epoch: 20/20... Training loss: 0.089
Epoch: 20/20... Training loss: 0.090
Epoch: 20/20... Training loss: 0.094
Epoch: 20/20... Training loss: 0.088
Epoch: 20/20... Training loss: 0.094
Epoch: 20/20... Training loss: 0.093
Epoch: 20/20... Training loss: 0.091
Epoch: 20/20... Training loss: 0.095
Epoch: 20/20... Training loss: 0.093
Epoch: 20/20... Training loss: 0.091
Epoch: 20/20... Training loss: 0.094
Epoch: 20/20... Training loss: 0.090
Epoch: 20/20... Training loss: 0.091
Epoch: 20/20... Training loss: 0.095
Epoch: 20/20... Training loss: 0.095
Epoch: 20/20... Training loss: 0.094
Epoch: 20/20... Training loss: 0.092
Epoch: 20/20... Training loss: 0.092
Epoch: 20/20... Training loss: 0.093
Epoch: 20/20... Training loss: 0.093
```

在运行前面的代码段后，网络已经训练了 20 次，我们将获得一个训练好的模型，该模型能够从 MNIST 数据集的测试集中生成或重建图像。请记住，如果我们提供的图像与训练模型的图像不相似，那么重建过程将无法进行，因为自动编码器只能用于特定的数据。

接下来,通过从测试集中抽取一些图像来测试训练模型,并查看模型是否能在解码器部分中重建它们。

```
fig, axes = plt.subplots(nrows=2, ncols=10, sharex=True, sharey=True,
figsize=(20,4))

input_images = mnist_dataset.test.images[:10]
reconstructed_images, compressed_images = sess.run([decoding_layer,
encoding_layer], feed_dict={inputs_values: input_images})

for imgs, row in zip([input_images, reconstructed_images], axes):
    for img, ax in zip(imgs, row):
        ax.imshow(img.reshape((28, 28)), cmap='Greys_r')
        ax.get_xaxis().set_visible(False)
        ax.get_yaxis().set_visible(False)
fig.tight_layout(pad=0.1)
```

输出如图 13.7 所示。

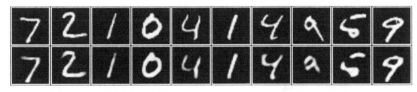

图 13.7 原始测试集中的图像(第一行)及其重构的样本(第二行)

在图 13.7 中,重建的图像非常接近输入图像,但我们可以通过在编码器-解码器部分中使用卷积层获得更好的图像。

13.5 卷积自动编码器

之前的简单实现在尝试从 MNIST 数据集中重建输入图像时做得很好,但我们可以通过在自动编码器的编码器和解码器中使用卷积层来获得更好的表现。替换的结果网络称为**卷积自动编码器**(convolutional autoencoder,CAE)。这种能够替换部分层的灵活性是自动编码器的一大优势,并使它们适用于不同的领域。

用于 CAE 的架构在网络的解码器部分中包含上采样层,以获得图像的重建版本。

13.5.1 数据集

在此实现中,我们可以使用任何类型的图像数据集,并查看自动编码器的卷积版本

第 13 章 自动编码器——特征提取和降噪

将如何发挥作用。因为这次仍将使用 MNIST 数据集，所以现在从使用 TensorFlow 辅助函数获取数据集开始。

```
%matplotlib inline

import numpy as np
import tensorflow as tf
import matplotlib.pyplot as plt

from tensorflow.examples.tutorials.mnist import input_data
mnist_dataset = input_data.read_data_sets('MNIST_data', validation_size=0)
```

输出如下。

```
from tensorflow.examples.tutorials.mnist import input_data

mnist_dataset = input_data.read_data_sets('MNIST_data', validation_size=0)

Extracting MNIST_data/train-images-idx3-ubyte.gz
Extracting MNIST_data/train-labels-idx1-ubyte.gz
Extracting MNIST_data/t10k-images-idx3-ubyte.gz
Extracting MNIST_data/t10k-labels-idx1-ubyte.gz
```

现在从数据集中显示一个数字。

```
# Plotting one image from the training set.
image = mnist_dataset.train.images[2]
plt.imshow(image.reshape((28, 28)), cmap='Greys_r')
```

输出结果如图 13.8 所示。

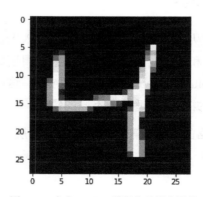

图 13.8　来自 MNIST 数据集的样本图像

13.5.2 构建模型

在该实现中,我们将使用步长(stride)为 1 的卷积层,并且将 **padding** 参数设置为 same,这样就不会改变图像的高度或宽度。此外,这里使用一组最大池化层来减小图像的宽度和高度,从而构建图像的压缩低维表示。

接下来继续构建网络的核心。

```
learning_rate = 0.001

# Define the placeholder variable sfor the input and target values
inputs_values = tf.placeholder(tf.float32, (None, 28,28,1),
name="inputs_values")
targets_values = tf.placeholder(tf.float32, (None, 28,28,1),
name="targets_values")

# Defining the Encoder part of the netowrk
# Defining the first convolution layer in the encoder parrt
# The output tenosor will be in the shape of 28x28x16
conv_layer_1 = tf.layers.conv2d(inputs=inputs_values, filters=16,
kernel_size=(3,3), padding='same', activation=tf.nn.relu)

# The output tenosor will be in the shape of 14x14x16
maxpool_layer_1 = tf.layers.max_pooling2d(conv_layer_1, pool_size=(2,2),
strides=(2,2), padding='same')

# The output tenosor will be in the shape of 14x14x8
conv_layer_2 = tf.layers.conv2d(inputs=maxpool_layer_1, filters=8,
kernel_size=(3,3), padding='same', activation=tf.nn.relu)

# The output tenosor will be in the shape of 7x7x8
maxpool_layer_2 = tf.layers.max_pooling2d(conv_layer_2, pool_size=(2,2),
strides=(2,2), padding='same')

# The output tenosor will be in the shape of 7x7x8
conv_layer_3 = tf.layers.conv2d(inputs=maxpool_layer_2, filters=8,
kernel_size=(3,3), padding='same', activation=tf.nn.relu)

# The output tenosor will be in the shape of 4x4x8
encoded_layer = tf.layers.max_pooling2d(conv_layer_3, pool_size=(2,2),
```

```
                strides=(2,2), padding='same')

# Defining the Decoder part of the netowrk
# Defining the first upsampling layer in the decoder part
# The output tenosor will be in the shape of 7x7x8
upsample_layer_1 = tf.image.resize_images(encoded_layer, size=(7,7),
method=tf.image.ResizeMethod.NEAREST_NEIGHBOR)

# The output tenosor will be in the shape of 7x7x8
conv_layer_4 = tf.layers.conv2d(inputs=upsample_layer_1, filters=8,
kernel_size=(3,3), padding='same', activation=tf.nn.relu)

# The output tenosor will be in the shape of 14x14x8
upsample_layer_2 = tf.image.resize_images(conv_layer_4, size=(14,14),
method=tf.image.ResizeMethod.NEAREST_NEIGHBOR)

# The output tenosor will be in the shape of 14x14x8
conv_layer_5 = tf.layers.conv2d(inputs=upsample_layer_2, filters=8,
kernel_size=(3,3), padding='same', activation=tf.nn.relu)

# The output tenosor will be in the shape of 28x28x8
upsample_layer_3 = tf.image.resize_images(conv_layer_5, size=(28,28),
method=tf.image.ResizeMethod.NEAREST_NEIGHBOR)

# The output tenosor will be in the shape of 28x28x16
conv6 = tf.layers.conv2d(inputs=upsample_layer_3, filters=16,
kernel_size=(3,3), padding='same', activation=tf.nn.relu)

# The output tenosor will be in the shape of 28x28x1
logits_layer = tf.layers.conv2d(inputs=conv6, filters=1, kernel_size=(3,3),
padding='same', activation=None)

# feeding the logits values to the sigmoid activation function to get the
reconstructed images
decoded_layer = tf.nn.sigmoid(logits_layer)

# feeding the logits to sigmoid while calculating the cross entropy
model_loss = tf.nn.sigmoid_cross_entropy_with_logits(labels=targets_values,
logits=logits_layer)
```

```
# Getting the model cost and defining the optimizer to minimize it
model_cost = tf.reduce_mean(model_loss)
model_optimizer =
tf.train.AdamOptimizer(learning_rate).minimize(model_cost)
```

令人高兴的是，我们构建了包含卷积神经网的解码器-解码器部分，同时展示了如何在解码器部分重建具有相同维度的输入图像。

13.5.3 训练模型

既然我们已经建立了模型，就可以通过从 MNIST 数据集生成随机批次数据，并将它们提供给先前定义的优化器来进入学习阶段。

现在从创建会话变量开始，它将负责执行之前定义的计算图。

```
sess = tf.Session()
num_epochs = 20
train_batch_size = 200
sess.run(tf.global_variables_initializer())

for e in range(num_epochs):
    for ii in range(mnist_dataset.train.num_examples//train_batch_size):
        input_batch = mnist_dataset.train.next_batch(train_batch_size)
        input_images = input_batch[0].reshape((-1, 28, 28, 1))
        input_batch_cost, _ = sess.run([model_cost, model_optimizer],
feed_dict={inputs_values: input_images,targets_values: input_images})

        print("Epoch: {}/{}...".format(e+1, num_epochs),
              "Training loss: {:.3f}".format(input_batch_cost))
```

输出如下。

```
.
.
.
Epoch: 20/20... Training loss: 0.102
Epoch: 20/20... Training loss: 0.099
Epoch: 20/20... Training loss: 0.103
Epoch: 20/20... Training loss: 0.102
```

```
Epoch: 20/20... Training loss: 0.100
Epoch: 20/20... Training loss: 0.101
Epoch: 20/20... Training loss: 0.098
Epoch: 20/20... Training loss: 0.103
Epoch: 20/20... Training loss: 0.104
Epoch: 20/20... Training loss: 0.103
Epoch: 20/20... Training loss: 0.098
Epoch: 20/20... Training loss: 0.102
Epoch: 20/20... Training loss: 0.098
Epoch: 20/20... Training loss: 0.099
Epoch: 20/20... Training loss: 0.103
Epoch: 20/20... Training loss: 0.104
Epoch: 20/20... Training loss: 0.101
Epoch: 20/20... Training loss: 0.105
Epoch: 20/20... Training loss: 0.102
Epoch: 20/20... Training loss: 0.102
Epoch: 20/20... Training loss: 0.100
Epoch: 20/20... Training loss: 0.099
Epoch: 20/20... Training loss: 0.102
Epoch: 20/20... Training loss: 0.102
Epoch: 20/20... Training loss: 0.104
Epoch: 20/20... Training loss: 0.101
Epoch: 20/20... Training loss: 0.099
Epoch: 20/20... Training loss: 0.098
Epoch: 20/20... Training loss: 0.100
Epoch: 20/20... Training loss: 0.101
Epoch: 20/20... Training loss: 0.100
Epoch: 20/20... Training loss: 0.100
Epoch: 20/20... Training loss: 0.101
Epoch: 20/20... Training loss: 0.098
Epoch: 20/20... Training loss: 0.101
Epoch: 20/20... Training loss: 0.103
Epoch: 20/20... Training loss: 0.103
Epoch: 20/20... Training loss: 0.102
Epoch: 20/20... Training loss: 0.101
Epoch: 20/20... Training loss: 0.100
Epoch: 20/20... Training loss: 0.101
Epoch: 20/20... Training loss: 0.102
Epoch: 20/20... Training loss: 0.103
```

```
Epoch: 20/20... Training loss: 0.103
Epoch: 20/20... Training loss: 0.103
Epoch: 20/20... Training loss: 0.099
Epoch: 20/20... Training loss: 0.101
Epoch: 20/20... Training loss: 0.096
Epoch: 20/20... Training loss: 0.104
Epoch: 20/20... Training loss: 0.104
Epoch: 20/20... Training loss: 0.103
Epoch: 20/20... Training loss: 0.103
Epoch: 20/20... Training loss: 0.104
Epoch: 20/20... Training loss: 0.099
Epoch: 20/20... Training loss: 0.101
Epoch: 20/20... Training loss: 0.101
Epoch: 20/20... Training loss: 0.099
Epoch: 20/20... Training loss: 0.100
Epoch: 20/20... Training loss: 0.102
Epoch: 20/20... Training loss: 0.100
Epoch: 20/20... Training loss: 0.098
Epoch: 20/20... Training loss: 0.100
Epoch: 20/20... Training loss: 0.097
Epoch: 20/20... Training loss: 0.102
```

运行前面的代码段后，网络已经被训练了 20 次，我们将获得一个训练好的 CAE，因此现在通过从 MNIST 数据集中抽取类似的图像来测试这个模型。

```
fig, axes = plt.subplots(nrows=2, ncols=10, sharex=True, sharey=True,
figsize=(20,4))
input_images = mnist_dataset.test.images[:10]
reconstructed_images = sess.run(decoded_layer, feed_dict={inputs_values:
input_images.reshape((10, 28, 28, 1))})

for imgs, row in zip([input_images, reconstructed_images], axes):
    for img, ax in zip(imgs, row):
        ax.imshow(img.reshape((28, 28)), cmap='Greys_r')
        ax.get_xaxis().set_visible(False)
        ax.get_yaxis().set_visible(False)
fig.tight_layout(pad=0.1)
```

输出如图 13.9 所示。

图 13.9 原始测试图像（第一行）及使用
卷积自动编码器重构的样本（第二行）

13.6 降噪自动编码器

我们可以通过强制自动编码器学习有关输入数据的更重要的特征，来进一步增强自动编码器的性能。通过向输入图像添加噪声，并以原始噪声作为目标，模型将尝试消除此噪声并学习有关它们的重要特征，以便输出有意义的重构图像。这种 CAE 架构可用于从输入图像中去除噪声。自动编码器的这种特定变体称为**降噪自动编码器**（denoising autoencoder），如图 13.10 所示。

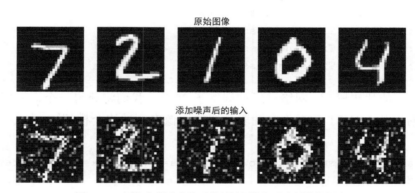

图 13.10 添加高斯噪声前后的原始图像和噪声图像的样本

下面从实现图 13.11 中的架构来开始完成这个任务。在这种降噪自动编码器架构中额外添加的唯一东西是原始输入图像中的一些噪声。

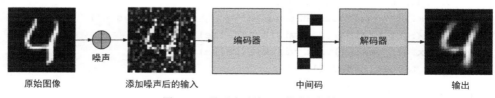

图 13.11 降噪自动编码器的常规架构

13.6.1 构建模型

在此实现中,我们将在编码器和解码器部分中使用具有更多层的神经网络,其原因是现在任务的复杂性已经提高。

模型剩下的部分与之前的 CAE 完全相同,但添加了额外的层来帮助模型从嘈杂的图像中重建无噪声图像。

接下来,继续构建这个模型。

```
learning_rate = 0.001

# Define the placeholder variable sfor the input and target values
inputs_values = tf.placeholder(tf.float32, (None, 28, 28, 1),
name='inputs_values')
targets_values = tf.placeholder(tf.float32, (None, 28, 28, 1),
name='targets_values')

# Defining the Encoder part of the netowrk
# Defining the first convolution layer in the encoder parrt
# The output tenosor will be in the shape of 28x28x32
conv_layer_1 = tf.layers.conv2d(inputs=inputs_values, filters=32,
kernel_size=(3,3), padding='same', activation=tf.nn.relu)

# The output tenosor will be in the shape of 14x14x32
maxpool_layer_1 = tf.layers.max_pooling2d(conv_layer_1, pool_size=(2,2),
strides=(2,2), padding='same')

# The output tenosor will be in the shape of 14x14x32
conv_layer_2 = tf.layers.conv2d(inputs=maxpool_layer_1, filters=32,
kernel_size=(3,3), padding='same', activation=tf.nn.relu)

# The output tenosor will be in the shape of 7x7x32
maxpool_layer_2 = tf.layers.max_pooling2d(conv_layer_2, pool_size=(2,2),
strides=(2,2), padding='same')

# The output tenosor will be in the shape of 7x7x16
conv_layer_3 = tf.layers.conv2d(inputs=maxpool_layer_2, filters=16,
kernel_size=(3,3), padding='same', activation=tf.nn.relu)

# The output tenosor will be in the shape of 4x4x16
```

```python
encoding_layer = tf.layers.max_pooling2d(conv_layer_3, pool_size=(2,2),
strides=(2,2), padding='same')

# Defining the Decoder part of the netowrk
# Defining the first upsampling layer in the decoder part
# The output tenosor will be in the shape of 7x7x16
upsample_layer_1 = tf.image.resize_images(encoding_layer, size=(7,7),
method=tf.image.ResizeMethod.NEAREST_NEIGHBOR)

# The output tenosor will be in the shape of 7x7x16
conv_layer_4 = tf.layers.conv2d(inputs=upsample_layer_1, filters=16,
kernel_size=(3,3), padding='same', activation=tf.nn.relu)

# The output tenosor will be in the shape of 14x14x16
upsample_layer_2 = tf.image.resize_images(conv_layer_4, size=(14,14),
method=tf.image.ResizeMethod.NEAREST_NEIGHBOR)

# The output tenosor will be in the shape of 14x14x32
conv_layer_5 = tf.layers.conv2d(inputs=upsample_layer_2, filters=32,
kernel_size=(3,3), padding='same', activation=tf.nn.relu)

# The output tenosor will be in the shape of 28x28x32
upsample_layer_3 = tf.image.resize_images(conv_layer_5, size=(28,28),
method=tf.image.ResizeMethod.NEAREST_NEIGHBOR)

# The output tenosor will be in the shape of 28x28x32
conv_layer_6 = tf.layers.conv2d(inputs=upsample_layer_3, filters=32,
kernel_size=(3,3), padding='same', activation=tf.nn.relu)

# The output tenosor will be in the shape of 28x28x1
logits_layer = tf.layers.conv2d(inputs=conv_layer_6, filters=1,
kernel_size=(3,3), padding='same', activation=None)

# feeding the logits values to the sigmoid activation function to get the
reconstructed images
decoding_layer = tf.nn.sigmoid(logits_layer)

# feeding the logits to sigmoid while calculating the cross entropy
model_loss = tf.nn.sigmoid_cross_entropy_with_logits(labels=targets_values,
logits=logits_layer)
```

```
# Getting the model cost and defining the optimizer to minimize it
model_cost = tf.reduce_mean(model_loss)
model_optimizer =
tf.train.AdamOptimizer(learning_rate).minimize(model_cost)
```

至此我们就构建成了一个更复杂（更深层次）的卷积模型。

13.6.2 训练模型

下面开始训练这个更深的网络，这也意味着将花费更多的时间来重构来自噪声输入的无噪声图像。

首先，创建会话变量。

```
sess = tf.Session()
```

然后，开始训练过程，但这次需要更多的训练次数。

```
num_epochs = 100
train_batch_size = 200

# Defining a noise factor to be added to MNIST dataset
mnist_noise_factor = 0.5
sess.run(tf.global_variables_initializer())

for e in range(num_epochs):
    for ii in range(mnist_dataset.train.num_examples//train_batch_size):
        input_batch = mnist_dataset.train.next_batch(train_batch_size)
        # Getting and reshape the images from the corresponding batch
        batch_images = input_batch[0].reshape((-1, 28, 28, 1))
        # Add random noise to the input images
        noisy_images = batch_images + mnist_noise_factor * np.random.randn(*batch_images.shape)
        # Clipping all the values that are above 0 or above 1
        noisy_images = np.clip(noisy_images, 0., 1.)
        # Set the input images to be the noisy ones and the original images to be the target
        input_batch_cost, _ = sess.run([model_cost, model_optimizer],
                feed_dict={inputs_values: noisy_images,
                                                        targets_values: batch_images})
```

```
            print("Epoch: {}/{}...".format(e+1, num_epochs),
                  "Training loss: {:.3f}".format(input_batch_cost))
```

输出如下。

.
.
.

```
Epoch: 100/100... Training loss: 0.098
Epoch: 100/100... Training loss: 0.101
Epoch: 100/100... Training loss: 0.103
Epoch: 100/100... Training loss: 0.098
Epoch: 100/100... Training loss: 0.102
Epoch: 100/100... Training loss: 0.102
Epoch: 100/100... Training loss: 0.103
Epoch: 100/100... Training loss: 0.101
Epoch: 100/100... Training loss: 0.098
Epoch: 100/100... Training loss: 0.099
Epoch: 100/100... Training loss: 0.096
Epoch: 100/100... Training loss: 0.100
Epoch: 100/100... Training loss: 0.100
Epoch: 100/100... Training loss: 0.103
Epoch: 100/100... Training loss: 0.100
Epoch: 100/100... Training loss: 0.101
Epoch: 100/100... Training loss: 0.099
Epoch: 100/100... Training loss: 0.096
Epoch: 100/100... Training loss: 0.102
Epoch: 100/100... Training loss: 0.099
Epoch: 100/100... Training loss: 0.098
Epoch: 100/100... Training loss: 0.102
Epoch: 100/100... Training loss: 0.100
Epoch: 100/100... Training loss: 0.100
Epoch: 100/100... Training loss: 0.099
Epoch: 100/100... Training loss: 0.098
Epoch: 100/100... Training loss: 0.100
Epoch: 100/100... Training loss: 0.099
Epoch: 100/100... Training loss: 0.102
Epoch: 100/100... Training loss: 0.099
Epoch: 100/100... Training loss: 0.102
Epoch: 100/100... Training loss: 0.100
Epoch: 100/100... Training loss: 0.101
Epoch: 100/100... Training loss: 0.102
Epoch: 100/100... Training loss: 0.098
```

```
Epoch: 100/100... Training loss: 0.103
Epoch: 100/100... Training loss: 0.100
Epoch: 100/100... Training loss: 0.098
Epoch: 100/100... Training loss: 0.100
Epoch: 100/100... Training loss: 0.097
Epoch: 100/100... Training loss: 0.099
Epoch: 100/100... Training loss: 0.100
Epoch: 100/100... Training loss: 0.101
Epoch: 100/100... Training loss: 0.101
```

此时训练的模型已经能够输出无噪声图像,这可以使得自动编码器适用于更多领域。

在下一段代码中,不会再将 MNIST 测试集的原始图像提供给模型,因为需要首先向这些图像添加噪声,以查看训练模型是否能够生成无噪声图像。

这里将噪声添加到测试图像中,并将它们传递给自动编码器。尽管有时肉眼很难分辨输出的原始数字是什么,但降噪自动编码器在消除噪声方面依然做得非常出色。

```
#Defining some figures
fig, axes = plt.subplots(nrows=2, ncols=10, sharex=True, sharey=True,
figsize=(20,4))

#Visualizing some images
input_images = mnist_dataset.test.images[:10]
noisy_imgs = input_images + mnist_noise_factor *
np.random.randn(*input_images.shape)

#Clipping and reshaping the noisy images
noisy_images = np.clip(noisy_images, 0., 1.).reshape((10, 28, 28, 1))

#Getting the reconstructed images
reconstructed_images = sess.run(decoding_layer, feed_dict={inputs_values:
noisy_images})

#Visualizing the input images and the noisy ones
for imgs, row in zip([noisy_images, reconstructed_images], axes):
    for img, ax in zip(imgs, row):
        ax.imshow(img.reshape((28, 28)), cmap='Greys_r')
        ax.get_xaxis().set_visible(False)
        ax.get_yaxis().set_visible(False)

fig.tight_layout(pad=0.1)
```

输出如图 13.12 所示。

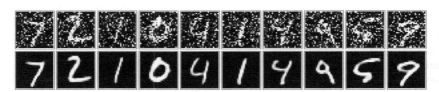

图 13.12 具有高斯噪声的原始测试图像（第一行）及
其基于已训练的降噪自动编码器重构后的样本

13.7 自动编码器的应用

在前文提到的从低维表示构建图像的示例中，我们可以看到它的输出与原始输入非常相似，之后也看到了 CAN 的优势，它在对噪声数据集进行降噪时非常有效。前面实现的这种模型对于图像重构和数据集降噪非常有用，因此我们可以将上述实现迁移到自己感兴趣的任何其他任务中。

此外，在本章中，我们已经看到了自动编码器架构的灵活性以及如何对它进行不同的更改。我们甚至对它进行了测试，以解决从输入图像中去除噪声这样的难题。自动编码器的这种灵活性使它能够解决更多的应用问题。

13.7.1 图像着色

自动编码器（尤其是卷积版）可用于较难的任务，如图像着色。在下面的示例中，将传递给模型没有任何颜色的输入图像，并且自动编码器模型将输出对此图像着色后的重构版本，如图 13.13 所示。

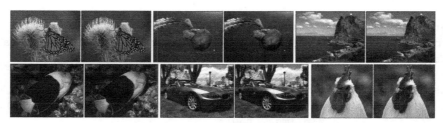

图 13.13 CAE 经过训练后用来对图像进行着色
（彩色版本见本书在异步社区中的配套资源）

这里，着色模型架构如图 13.14 所示。

至此，自动编码器已经过训练，我们可以用它来为以前从未见过的图片着色。

这种应用程序可用于为早期用相机拍摄的非常老的图片着色。

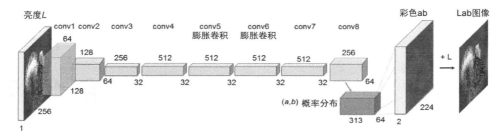

图 13.14 着色模型架构

13.7.2 更多的应用

另一个有趣的应用是生成具有更高分辨率的图像,这称为神经图像增强。

Richard Zhang 等的这些图片展示了一些更真实的着色图像,如图 13.15 所示。

图 13.15 Richard Zhang、Phillip Isola 和 Alexei A. Efros 实现的彩色图像着色
（彩色版本见本书在异步社区中的配套资源）

图 13.16 显示了自动编码器的另一个应用——增强图像。

图 13.16 Alexjc 的神经增强效果（图片来源：GitHub 网站,
彩色版本见本书在异步社区中的配套资源）

13.8 总结

本章介绍了一种可用于许多有趣任务的全新架构。自动编码器非常灵活,可以在图像增强、着色或重构方面提供有效的解决方案。此外,还有更多的自动编码器变体,称为**变分自动编码器**（variational autoencoder）。它们也有非常有趣的应用,例如图像生成。

第 14 章
生成对抗网络

生成对抗网络（Generative Adversarial Network，GAN）是一种由两个彼此对抗（因此有了名字中的**对抗**）的网络组成的深度神经网络结构。

2014 年，Ian Goodfellow 和包括 Yoshua Benjio 在内的其他研究人员在蒙特利尔大学的一篇论文"Generative Adversarial Networks"中介绍了 GAN。提到 GAN，Facebook 的 AI 研究总裁 Yann LeCun 称**对抗训练**是机器学习过去十年中最有趣的想法。

GAN 潜力巨大，因为它可以学习模仿任何的数据分布。也就是说，可以训练 GAN 在包括图像、音乐、语言和文本在内的任何领域中创造与我们类似的世界。从某种意义上说，它们是机器人艺术家，它们的输出令人印象深刻（见 nytimes 网站）同时也让人们受到鼓舞。

本章将主要包括以下几个主题。

- 直观介绍。
- GAN 的简单实现。
- 深度卷积 GAN。

14.1 直观介绍

本节将会以非常直观的方式介绍生成对抗网络 GAN。为了了解 GAN 是如何运作的，这里将采用一个获取活动门票的虚构场景。

故事从某个地方正在举行的非常有趣的活动开始，并且你很有兴趣参加。可你很晚才听说这个活动，所有门票都已售罄，但是你想方设法也要参加活动。所以你想出了一个主意。你想尝试伪造一张和原始门票完全相同或者非常相似的门票。但是生活并不是这么容易的，这里还有另一个挑战：你不知道原始门票是什么样的。因此，根据你参加此类活动的经验，你开始想象门票应该长什么样，并凭着想象开始设计门票。

你开始尝试设计门票，然后去参加活动并向保安人员出示门票，你希望他们能够相信你并且放你进去。但是你不想多次向保安人员展示你的面孔，所以你决定从朋友那里获取帮助，他们会带着你凭空设计的门票去参加聚会并向保安人员出示。如果保安人员不让他进去，他将会通过查看一些拿着真实门票入场的人，来获取一些关于门票外观的信息。你将会根据朋友的评论来调整门票，直到保安人员让他进去。此时，仅在此时，你将设计另一张完全相同的门票，然后自己拿着它入场。

不要过多考虑这个故事有多么不现实，事实上 GAN 的运行方式就和这个故事非常相似。GAN 如今非常流行，人们正在将它们用于计算机视觉领域的很多应用中。

有很多有趣的应用会使用到 GAN，这里将会提及并实现其中的一些。

GAN 中有两个主要组件，它们在众多计算机视觉领域取得了突破。第一个组件叫作**生成器**，第二个叫作**判别器**，如图 14.1 所示。

- 生成器将会尝试从某个特定概率分布中产生数据样本，这与前面提到的尝试复制活动门票的人非常相似。

- 判别器将会判别（就像前面试图通过在门票中找到漏洞，来判断门票真伪的安保人员）输入是来自于原始的训练集（原始门票），还是来自于生成器部分（被尝试复制原始门票的人设计出来的）。

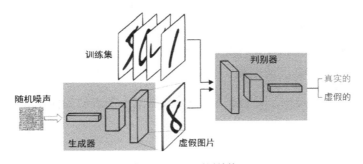

图 14.1　GAN 通用结构

14.2　GAN 的简单实现

从伪造活动门票的故事中，可以非常直观地看出 GAN 的思想。为了清楚地理解 GAN

是如何工作的以及如何实现它们，本节将会在 MNIST 数据集上演示 GAN 的一个简单实现。

首先，需要构建 GAN 网络的核心，它由两个主要部分组成：生成器和判别器。生成器将会尝试从某个特定的概率分布中想象或者伪造数据样本；而可以访问和查看实际数据样本的判别器将会判断生成器的输出是在设计中存在缺陷还是它与原始数据样本非常接近。与前面的活动场景相似，生成器的整个目的就是使得判别器相信生成的图像是来自真实数据集的，以此来试图欺骗判别器。

训练过程和前面的故事有着相似的结尾，生成器最终将会设法生成与原始数据样本看起来非常相似的图像。

图 14.2 显示了 GAN 的典型结构，将在 MNIST 数据集上训练 GAN。图 14.2 中的隐藏样本部分是一个随机想法或者向量，生成器将会使用它来从真实图像中复制出虚假图像。

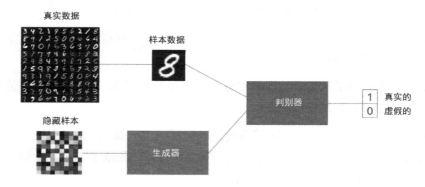

图 14.2 针对 MNIST 数据集的通用 GAN 架构

正如前文提到的，作为一个判断者，判别器将尝试从生成器设计的虚假图像中分辨出真实图像。所以这个网络将产生一个二值输出，二值输出可以使用 sigmoid 函数来表示（0 表示输入的是虚假图像，1 表示输入的是真实图像）。

现在继续实现这个架构，看它在 MNIST 数据集上的表现如何。

从导入此实现所需的库开始。

```
%matplotlib inline

import matplotlib.pyplot as plt
import pickle as pkl

import numpy as np
import tensorflow as tf
```

14.2 GAN 的简单实现

因为这里使用了 MNIST 数据集,所以将会使用 TensorFlow 辅助函数来获取数据集,并将它存储在某处。

```
from tensorflow.examples.tutorials.mnist import input_data
mnist_dataset = input_data.read_data_sets('MNIST_data')
```

输出如下。

```
Extracting MNIST_data/train-images-idx3-ubyte.gz
Extracting MNIST_data/train-labels-idx1-ubyte.gz
Extracting MNIST_data/t10k-images-idx3-ubyte.gz
Extracting MNIST_data/t10k-labels-idx1-ubyte.gz
```

14.2.1 模型输入

在深入构建由生成器和判别器表示的 GAN 的核心之前,先定义计算图的输入。如图 14.3 所示,需要两个输入:一个输入是真实图像,会把它提供给判别器;另一个输入称为**隐空间**,会将它提供给生成器,并用于生成虚假图像。

```
# Defining the model input for the generator and discriminator
def inputs_placeholders(discriminator_real_dim, gen_z_dim):
    real_discrminator_input = tf.placeholder(tf.float32, (None, 
discrimator_real_dim), name="real_discrminator_input")
    generator_inputs_z = tf.placeholder(tf.float32, (None, gen_z_dim), 
name="generator_input_z")
    return real_discrminator_input, generator_inputs_z
```

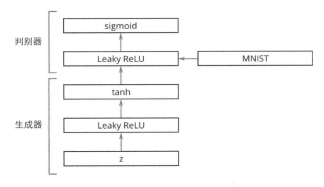

图 14.3 在 MNIST 数据集上实现的 GAN 架构

下面开始深入构建 GAN 架构的两个核心组件。首先从构建生成器部分开始。如图

14.3 所示，生成器将包含至少一个隐藏层，它将作为一个近似器。此外，将会采用一种称为 Leaky ReLU 的激活函数，而不是通用的 ReLU 激活函数。这将允许梯度值在层与层之间随意流动（关于 Leaky ReLU 的更多信息将会在 14.2.3 节中介绍）。

14.2.2 变量作用域

变量作用域是 TensorFlow 中的一个特性，作用域有助于执行如下操作。

- 确保有一些命名约定，以便后续检索变量。例如，通过使变量以单词 generator 或 discriminator 开头，这在网络训练期间将有所帮助。其实也可以使用名字作用域特性，但是这个特性不能帮助我们实现第二个目的。
- 能够重复使用或重复训练有不同输入的相同网络。例如，我们将从生成器中对虚假图像进行采样，来查看生成器复制原始图像的性能如何。此外，判别器可以访问真实图像和虚假图像，这使得在构建计算图时可以轻松地重用变量而不是创建新变量。

以下语句将说明如何使用 TensorFlow 中的变量作用域特性。

```
with tf.variable_scope('scopeName', reuse=False):
    # Write your code here
```

读者可以在 TensorFlow 官网中搜索"variable scope"来了解关于使用变量作用域特性的更多好处。

14.2.3 Leaky ReLU

前文提到，使用与 ReLU 激活函数不同版本的激活函数——Leaky ReLU。传统版本的 ReLU 激活函数通过其他方式将负值截断为零，只会取输入值和零值中的最大值。而这里使用的 Leaky ReLU 版本允许存在一些负值，因此得名 **Leaky ReLU**。

有时使用传统的 ReLU 激活函数，网络会陷入一种常态——死亡状态，这是因为网络所有的输出全为零。

Leaky ReLU 的思想是通过允许一些负值传递来阻止这种死亡状态。

使生成器工作的整个思想就是从判别器接收梯度值，并且如果网络陷入死亡状态，学习过程就不会出现。

图 14.4 和图 14.5 显示了传统 ReLU 与 Leaky ReLU 激活函数之间的不同。

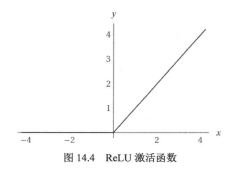

图 14.4　ReLU 激活函数

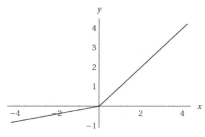

图 14.5　Leaky ReLU 激活函数

因为 Leaky ReLU 激活函数并没有在 TensorFlow 中实现，所以需要我们自己去实现它。如果输入为正数，此激活函数的输出也为正数；如果输入为负数，则此激活函数的输出将是一个受控制的负值。这里将使用一个称为 **alpha** 的参数来控制负值，通过允许传递一些负值来引入网络的容错性。

下面的等式表示需要实现的 Leaky ReLU 函数。

$$f(x) = \max(ax, x)$$

14.2.4　生成器

把 MNIST 图像归一化到 0～1，从而使得 sigmoid 激活函数充分发挥作用。但实际上，我们发现 tanh 激活函数比其他任何函数都具有更好的性能。因此为了使用 tanh 激活函数，需要将这些图像的像素值范围重新缩放到 −1～1。

```
def generator(gen_z, gen_out_dim, num_hiddern_units=128, reuse_vars=False,
leaky_relu_alpha=0.01):
    ''' Building the generator part of the network
        Function arguments
        ---------
        gen_z : the generator input tensor
        gen_out_dim : the output shape of the generator
        num_hiddern_units : Number of neurons/units in the hidden layer
        reuse_vars : Reuse variables with tf.variable_scope
        leaky_relu_alpha : leaky ReLU parameter
        Function Returns
        -------
        tanh_output, logits_layer:
    '''
    with tf.variable_scope('generator', reuse=reuse_vars):
```

```
        # Defining the generator hidden layer
        hidden_layer_1 = tf.layers.dense(gen_z, num_hiddern_units,
activation=None)
        # Feeding the output of hidden_layer_1 to leaky relu
        hidden_layer_1 = tf.maximum(hidden_layer_1,
leaky_relu_alpha*hidden_layer_1)
        # Getting the logits and tanh layer output
        logits_layer = tf.layers.dense(hidden_layer_1, gen_out_dim,
activation=None)
        tanh_output = tf.nn.tanh(logits_layer)
        return tanh_output, logits_layer
```

现在我们已经准备好了生成器部分，下面继续定义 GAN 的第二个组件。

14.2.5 判别器

接下来，构建生成对抗网络中的第二个主要组件，即判别器。判别器与生成器基本相同，但不使用 tanh 激活函数，而使用 sigmoid 激活函数；它将产生一个二值输出，代表判别器对输入图像的判断。

```
def discriminator(disc_input, num_hiddern_units=128, reuse_vars=False,
leaky_relu_alpha=0.01):
    ''' Building the discriminator part of the network
        Function Arguments
        ---------
        disc_input : discrminator input tensor
        num_hiddern_units : Number of neurons/units in the hidden layer
        reuse_vars : Reuse variables with tf.variable_scope
        leaky_relu_alpha : leaky ReLU parameter
        Function Returns
        -------
        sigmoid_out, logits_layer:
    '''
    with tf.variable_scope('discriminator', reuse=reuse_vars):
        # Defining the generator hidden layer
        hidden_layer_1 = tf.layers.dense(disc_input, num_hiddern_units,
activation=None)
        # Feeding the output of hidden_layer_1 to leaky relu
        hidden_layer_1 = tf.maximum(hidden_layer_1,
leaky_relu_alpha*hidden_layer_1)
        logits_layer = tf.layers.dense(hidden_layer_1, 1, activation=None)
        sigmoid_out = tf.nn.sigmoid(logits_layer)
```

```
    return sigmoid_out, logits_layer
```

14.2.6 构建 GAN 网络

在定义了构建生成器和判别器组件的主要函数之后,下面将它们堆叠起来,然后为此实现定义模型损失和优化器。

1. 模型超参数

可以通过改变下面一组超参数来微调 GAN。

```
# size of discriminator input image
#28 by 28 will flattened to be 784
input_img_size = 784

# size of the generator latent vector
gen_z_size = 100

# number of hidden units for the generator and discriminator hidden layers
gen_hidden_size = 128
disc_hidden_size = 128

#leaky ReLU alpha parameter which controls the leak of the function
leaky_relu_alpha = 0.01

# smoothness of the label
label_smooth = 0.1
```

2. 定义生成器和判别器

在定义了用于生成虚假 MNIST 图像(看起来和真实图像基本相同)的 GAN 架构的两个主要组件之后,下面使用目前已经定义的函数来构建网络。构建网络将遵循以下步骤。

(1) 定义模型输入,输入包含两个变量。其中一个变量是真实图像,把它输入判别器,另一个变量是生成器用于复制原始图像的隐空间。

(2) 调用前面定义的生成器函数来构建网络的生成器部分。

(3) 调用前面定义的判别器函数来构建网络的判别器部分,但是这里会调用该函数两次。第一次调用针对真实数据,第二次调用针对生成器生成的虚假数据。

(4)通过重用变量来保持真实图像和虚假图像的权重是一样的。

```
tf.reset_default_graph()

# creating the input placeholders for the discrminator and
generator
real_discrminator_input, generator_input_z =
inputs_placeholders(input_img_size, gen_z_size)

# Create the generator network
gen_model, gen_logits = generator(generator_input_z,
input_img_size, gen_hidden_size, reuse_vars=False,
leaky_relu_alpha=leaky_relu_alpha)

# gen_model is the output of the generator
# Create the generator network
disc_model_real, disc_logits_real =
discriminator(real_discrminator_input, disc_hidden_size,
reuse_vars=False, leaky_relu_alpha=leaky_relu_alpha)
disc_model_fake, disc_logits_fake = discriminator(gen_model,
disc_hidden_size, reuse_vars=True,
leaky_relu_alpha=leaky_relu_alpha)
```

3. 判别器与生成器损失

这一部分需要定义判别器和生成器损失,可以认为这是此实现中最富有技巧的部分。

我们知道生成器试图伪造原始图像,并且判别器作为判断者,同时接收来自生成器和原始输入的图像。因此在为每一部分设计损失时,需要关注两件事。

首先,网络的判别器部分要能够区分由生成器生成的虚假图像和来自原始训练样本的真实图像。在训练时,将给判别器部分提供一批分为两类的数据。第一类是来自原始输入的图像,第二类是生成器生成的虚假图像。

因此,判别器最终的总损失将是它接受真实图像为真实图像并且检测假图像为虚假图像的能力之和。最终的总损失如下。

$$disc\_loss = disc\_loss\_real + disc\_loss\_fake$$

```
tf.reduce_mean(tf.nn.sigmoid_cross_entropy_with_logits(logits=logits_layer,
labels=labels))
```

然后，需要计算两个损失才能得到最终的判别器损失。

第一个损失 disc_loss_real 将会根据从判别器和 labels 获得的 logits 值计算出来。在这种情况下，labels 的值全都为 1，因为此时最小批次中所有图像都来自 MNIST 数据集中的真实输入图像。为了增强模型在测试集上的泛化能力并给出更好的结果，我们发现其实将 labels 的值从 1 改为 0.9 会更好。标签的这种改变称为**标签平滑**。

```
labels = tf.ones_like(tensor) * (1 - smooth)
```

判别器损失的第二部分是判别器能够检测虚假图像的能力，损失介于从判别器获得的 logits 值和 labels 值之间。此时，所有的 labels 值都是零，因为已知这个最小批次中的所有图像都来自生成器，而不是来自原始输入。

既然已经讨论了判别器损失，那么同样也需要计算生成器损失。生成器损失称为 gen_loss，它介于 disc_logits_fake（判别器对于虚假图像的输出）和标签（全都为 1，因为生成器试图使判别器相信它生成的虚假图像）之间。

```
# calculating the losses of the discrimnator and generator
disc_labels_real = tf.ones_like(disc_logits_real) * (1 - label_smooth)
disc_labels_fake = tf.zeros_like(disc_logits_fake)

disc_loss_real =
tf.nn.sigmoid_cross_entropy_with_logits(labels=disc_labels_real,
logits=disc_logits_real)
disc_loss_fake =
tf.nn.sigmoid_cross_entropy_with_logits(labels=disc_labels_fake,
logits=disc_logits_fake)

#averaging the disc loss
disc_loss = tf.reduce_mean(disc_loss_real + disc_loss_fake)

#averaging the gen loss
gen_loss = tf.reduce_mean(
    tf.nn.sigmoid_cross_entropy_with_logits(
        labels=tf.ones_like(disc_logits_fake),
        logits=disc_logits_fake))
```

4. 优化器

最后是优化器部分。在本节中，将会定义训练过程中使用的优化标准。首先，将分别更新生成器和判别器的变量,因此需要能够检索每一部分的变量。

对于第一个优化器(即生成器 1),将从计算图中的可训练变量中检索以 generator 名称开头的所有变量,然后通过参考其名称来检查每个变量属于哪一模块。

同样也要对判别器的变量做同样的操作,方法是令其所有变量都以 discriminator 开头。在这之后,就可以将想要优化的变量列表传递给优化器。

TensorFlow 的变量作用域特性使得我们能够检索以某个字符串开头的变量,然后会有两个不同的变量列表,一个用于生成器,另一个用于判别器。

```
# building the model optimizer

learning_rate = 0.002

# Getting the trainable_variables of the computational graph, split into
Generator and Discrimnator parts
trainable_vars = tf.trainable_variables()
gen_vars = [var for var in trainable_vars if
var.name.startswith("generator")]
disc_vars = [var for var in trainable_vars if
var.name.startswith("discriminator")]

disc_train_optimizer = tf.train.AdamOptimizer().minimize(disc_loss,
var_list=disc_vars)
gen_train_optimizer = tf.train.AdamOptimizer().minimize(gen_loss,
var_list=gen_vars)
```

14.2.7 训练模型

下面开始训练过程,来看 GAN 如何设法生成类似于 MNIST 数据集中图像的图像。

```
train_batch_size = 100
num_epochs = 100
generated_samples = []
model_losses = []

saver = tf.train.Saver(var_list = gen_vars)

with tf.Session() as sess:
    sess.run(tf.global_variables_initializer())
```

```python
for e in range(num_epochs):
    for ii in range(mnist_dataset.train.num_examples//train_batch_size):
        input_batch = mnist_dataset.train.next_batch(train_batch_size)
        # Get images, reshape and rescale to pass to D
        input_batch_images = input_batch[0].reshape((train_batch_size,784))
        input_batch_images = input_batch_images*2 - 1
        # Sample random noise for G
        gen_batch_z = np.random.uniform(-1, 1, size=(train_batch_size, gen_z_size))
        # Run optimizers
        _ = sess.run(disc_train_optimizer, feed_dict={real_discrminator_input: input_batch_images, generator_input_z: gen_batch_z})
        _ = sess.run(gen_train_optimizer, feed_dict={generator_input_z: gen_batch_z})
    # At the end of each epoch, get the losses and print them out
    train_loss_disc = sess.run(disc_loss, {generator_input_z: gen_batch_z, real_discrminator_input: input_batch_images})
    train_loss_gen = gen_loss.eval({generator_input_z: gen_batch_z})
    print("Epoch {}/{}...".format(e+1, num_epochs),
          "Disc Loss: {:.3f}...".format(train_loss_disc),
          "Gen Loss: {:.3f}".format(train_loss_gen))
    # Save losses to view after training
    model_losses.append((train_loss_disc, train_loss_gen))
    # Sample from generator as we're training for viegenerator_inputs_zwing afterwards
    gen_sample_z = np.random.uniform(-1, 1, size=(16, gen_z_size))
    generator_samples = sess.run(
                    generator(generator_input_z, input_img_size, reuse_vars=True),
                    feed_dict={generator_input_z: gen_sample_z})
    generated_samples.append(generator_samples)
    saver.save(sess, './checkpoints/generator_ck.ckpt')

# Save training generator samples
with open('train_generator_samples.pkl', 'wb') as f:
    pkl.dump(generated_samples, f)
```

输出如下。

.
.
.
```
Epoch 71/100... Disc Loss: 1.078... Gen Loss: 1.361
Epoch 72/100... Disc Loss: 1.037... Gen Loss: 1.555
Epoch 73/100... Disc Loss: 1.194... Gen Loss: 1.297
Epoch 74/100... Disc Loss: 1.120... Gen Loss: 1.730
Epoch 75/100... Disc Loss: 1.184... Gen Loss: 1.425
Epoch 76/100... Disc Loss: 1.054... Gen Loss: 1.534
Epoch 77/100... Disc Loss: 1.457... Gen Loss: 0.971
Epoch 78/100... Disc Loss: 0.973... Gen Loss: 1.688
Epoch 79/100... Disc Loss: 1.324... Gen Loss: 1.370
Epoch 80/100... Disc Loss: 1.178... Gen Loss: 1.710
Epoch 81/100... Disc Loss: 1.070... Gen Loss: 1.649
Epoch 82/100... Disc Loss: 1.070... Gen Loss: 1.530
Epoch 83/100... Disc Loss: 1.117... Gen Loss: 1.705
Epoch 84/100... Disc Loss: 1.042... Gen Loss: 2.210
Epoch 85/100... Disc Loss: 1.152... Gen Loss: 1.260
Epoch 86/100... Disc Loss: 1.327... Gen Loss: 1.312
Epoch 87/100... Disc Loss: 1.069... Gen Loss: 1.759
Epoch 88/100... Disc Loss: 1.001... Gen Loss: 1.400
Epoch 89/100... Disc Loss: 1.215... Gen Loss: 1.448
Epoch 90/100... Disc Loss: 1.108... Gen Loss: 1.342
Epoch 91/100... Disc Loss: 1.227... Gen Loss: 1.468
Epoch 92/100... Disc Loss: 1.190... Gen Loss: 1.328
Epoch 93/100... Disc Loss: 0.869... Gen Loss: 1.857
Epoch 94/100... Disc Loss: 0.946... Gen Loss: 1.740
Epoch 95/100... Disc Loss: 0.925... Gen Loss: 1.708
Epoch 96/100... Disc Loss: 1.067... Gen Loss: 1.427
Epoch 97/100... Disc Loss: 1.099... Gen Loss: 1.573
Epoch 98/100... Disc Loss: 0.972... Gen Loss: 1.884
Epoch 99/100... Disc Loss: 1.292... Gen Loss: 1.610
Epoch 100/100... Disc Loss: 1.103... Gen Loss: 1.736
```

在模型迭代了 100 轮之后，得到一个训练好的模型，该模型能够产生类似于输入判别器中的原始输入图像。

```
fig, ax = plt.subplots()
model_losses = np.array(model_losses)
plt.plot(model_losses.T[0], label='Disc loss')
plt.plot(model_losses.T[1], label='Gen loss')
plt.title("Model Losses")
plt.legend()
```

输出如图 14.6 所示，可以看到判别器和生成器曲线对应的模型损失都在收敛。

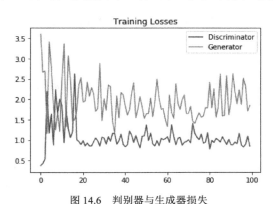

图 14.6　判别器与生成器损失

训练的生成器样本

下面测试模型的性能，并观察在训练过程将要结束时，生成器的生成能力（对应 14.1 节介绍的设计活动门票情景）如何得到增强。

```
def view_generated_samples(epoch_num, g_samples):
    fig, axes = plt.subplots(figsize=(7,7), nrows=4, ncols=4, sharey=True, sharex=True)
    print(gen_samples[epoch_num][1].shape)
    for ax, gen_image in zip(axes.flatten(), g_samples[0][epoch_num]):
        ax.xaxis.set_visible(False)
        ax.yaxis.set_visible(False)
        img = ax.imshow(gen_image.reshape((28,28)), cmap='Greys_r')
    return fig, axes
```

在绘制训练过程的最后一轮训练中生成的图像之前，需要加载持久化文件，该文件包含整个训练过程中每一轮训练生成的样本。

```
# Load samples from generator taken while training
with open('train_generator_samples.pkl', 'rb') as f:
    gen_samples = pkl.load(f)
```

现在我们绘制出了训练过程中最后一轮生成的 16 幅图像，下面来看生成器如何生成有意义的数字，如 3、7 和 2 等。

```
_ = view_generated_samples(-1, gen_samples)
```

输出结果如图 14.7 所示。

我们甚至可以查看生成器在不同轮的训练中的设计技巧。这里将显示生成器每 10 轮中生成的图像。

```
rows, cols = 10, 6
fig, axes = plt.subplots(figsize=(7,12), nrows=rows, ncols=cols,
sharex=True, sharey=True)

for gen_sample, ax_row in zip(gen_samples[::int(len(gen_samples)/rows)],
axes):
    for image, ax in zip(gen_sample[::int(len(gen_sample)/cols)], ax_row):
        ax.imshow(image.reshape((28,28)), cmap='Greys_r')
        ax.xaxis.set_visible(False)
        ax.yaxis.set_visible(False)
```

如图 14.8 所示，生成器的设计技巧及其生成虚假图像的能力起初都非常有限，在训练快结束时能力得到了增强。

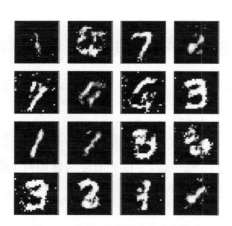

图 14.7　最后一轮训练中生成的样本

图 14.8　网络训练过程中每 10 轮生成的图像

14.2.8 从生成器中采样

14.2.7 节介绍了在 GAN 架构训练过程中生成的一些样本。我们还可以通过加载保存的检查点，并提供给生成器一个能够用于生成新图像的新隐空间变量，来从生成器产生一些全新的图像。

```
# Sampling from the generator
saver = tf.train.Saver(var_list=g_vars)

with tf.Session() as sess:
    #restoring the saved checkpints
    saver.restore(sess, tf.train.latest_checkpoint('checkpoints'))
    gen_sample_z = np.random.uniform(-1, 1, size=(16, z_size))
    generated_samples = sess.run(
                    generator(generator_input_z, input_img_size,
reuse_vars=True),
                    feed_dict={generator_input_z: gen_sample_z})
view_generated_samples(0, [generated_samples])
```

输出如图 14.9 所示。

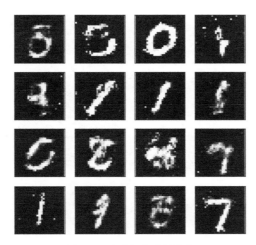

图 14.9　生成器生成的样本

在实现此示例时，有一些观测结果。在训练过程的第一轮迭代中，生成器无法产生和真实图像相似的图像，因为它根本不知道真实图像是什么样的。甚至判别器也不知道如何区分由生成器生成的虚假图像和真实图像。在训练开始时，出现了两种有趣的现象：

生成器不知道如何产生最初馈送到网络的真实图像，判别器也不知道真实图像和虚假图像之间的差别。

后来，生成器开始伪造在某种程度上有意义的图像，这是因为生成器将会学习到原始输入图像来自数据分布。同时，判别器将能够区分虚假图像与真实图像，但是它在训练过程接近尾声时将会被生成器欺骗。

14.3 总结

GAN 如今有许多有趣的应用，它可以用于不同的场景，如半监督和无监督任务。此外，由于大量研究人员致力于研究 GAN 网络，因此这些模型正在日益发展，它们生成图像和视频的能力也在变得越来越好。

这类模型可用于许多有趣的商业应用，例如，向 Photoshop 软件添加插件，该插件可以接受命令"make my smile more appealing"。同时，这类模型还可以用于图像降噪。

第 15 章 面部生成与标签缺失处理

使用 GAN 的有趣应用程序还有很多。本章将展示 GAN 的另一个很有前景的应用，即基于 CelebA 数据库的面部生成。同时还将演示当面对一个带有一些缺失标签的数据集时，如何将 GAN 用于半监督学习。

本章将介绍以下主题。
- 面部生成。
- 用生成对抗网络进行半监督学习。

15.1 面部生成

生成器与判别器包含了反卷积网络（Deconvolutional Network，DNN）[1]和卷积神经网络（Convolutional Neural Network，CNN）[2]。

- CNN 是一类神经网络，它将图像的数百个像素编码成维度更小的向量（z），可以认为这个向量是图像的压缩版本。
- DNN 是一类通过学习一些过滤器来从 z 恢复原始图像的网络。

此外，判别器将输出一个 1 或 0 来表示输入图像是来自实际数据集还是由生成器生成的。生成器将尝试基于隐空间 z 来生成类似于原始数据集的图像，隐空间 z 可能遵循高斯分布。因此，判别器的目标是正确识别真实图像，而生成器的目标是学习原始数据集的分布，从而欺骗判别器以使它做出错误的决定。

---

[1] 关于 DNN，详见 quora 官网。
[2] 关于 CNN，详见 CS231n 官网。

第 15 章 面部生成与标签缺失处理

在本节中，我们将尝试训练生成器学习人脸图像的分布，以便生成真实的面部图像。

对于大多数在应用中使用人类面孔的图形公司而言，生成类似人脸的面孔至关重要，同时它为我们提供了一个生动的例子，即人工智能如何在生成人类面孔时实现真实性。

这个示例将使用 CelebA 数据集。CelebFaces 属性数据集（CelebA）是一个大型的人脸属性数据集，有大约 20 万张名人图像，每张图像有 40 个属性注释。数据集中的图像涵盖了许多不同的姿势，同时背景也非常杂乱，CelebA 非常多样化并且有很好的注释。具体来说，其中包括以下内容。

- 10 177 个人。
- 202 599 张脸部图片。
- 每张图像的 5 个标志位置和 40 个二值属性注释。

除了面部生成之外，我们还可以将此数据集用于其他的计算机视觉应用，如面部识别和定位，或面部属性检测。

图 15.1 所示是训练过程中生成器的误差以及学习真实人脸分布的过程。

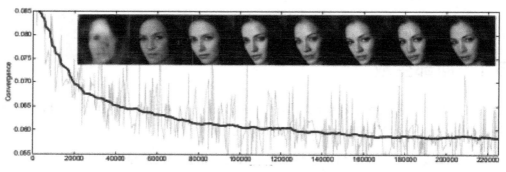

图 15.1 用于从名人图像数据集中生成新面孔的 GAN

15.1.1 获取数据

在本节中，将定义一些有助于下载 CelebA 数据集的辅助函数。首先，导入此实现所需的包。

```
import math
import os
import hashlib
from urllib.request import urlretrieve
```

```
import zipfile
import gzip
import shutil

import numpy as np
from PIL import Image
from tqdm import tqdm
import utils

import tensorflow as tf
```

然后，将使用 utils 脚本下载数据集。

```
#Downloading celebA dataset
celebA_data_dir = 'input'
utils.download_extract('celeba', celebA_data_dir)
```

输出如下。

```
Downloading celeba: 1.44GB [00:21, 66.6MB/s]
Extracting celeba...
```

15.1.2 探讨数据集

CelebA 数据集包含超过 20 万张带注释的名人图像。由于我们将使用 GAN 生成类似的图像，因此有必要查看该数据集中的一组图像并观察它们的外观。在本节中，将定义一些辅助函数，用于可视化 CelebA 数据集中的一组图像。

下面使用 utils 脚本来显示数据集中的一些图像。

```
#number of images to display
num_images_to_show = 25

celebA_images=utils.get_batch(glob(os.path.join(celebA_data_dir,
'img_align_celeba/*.jpg'))[:num_images_to_show],
                              28, 28, 'RGB')
pyplot.imshow(utils.images_square_grid(celebA_images, 'RGB'))
```

输出结果如图 15.2 所示。

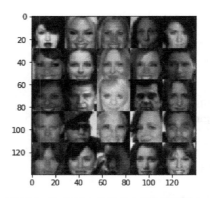

图 15.2 从 CelebA 数据集中提取图像样本

这个计算机视觉任务的主要焦点是使用 GAN 生成类似于名人数据集中的图像，因此我们需要关注图像的面部部分。要关注图像的脸部，就必须删除不包含名人脸部的图像部分。

15.1.3 构建模型

下面开始构建这次实现的核心，即计算图，主要包括以下部分。

- 模型输入。
- 判别器。
- 生成器。
- 模型的损失函数。
- 模型的优化器。
- 模型的训练环节。

1. 模型输入

在本节中，我们将实现一个辅助函数，用来定义负责为模型计算图提供数据输入的占位符。

这个函数应该能够创建 3 类主要的占位符。

- 来自数据集的实际输入图像，其维度包括批量大小、输入图像宽度、输入图像高度、通道数。

- 隐空间 z，发生器使用它生成虚假图像。
- 学习率的占位符。

辅助函数将返回这 3 个输入占位符构成的元组。下面继续定义这个函数。

```
# defining the model inputs
def inputs(img_width, img_height, img_channels, latent_space_z_dim):
    true_inputs = tf.placeholder(tf.float32, (None, img_width, img_height, img_channels),
                                              'true_inputs')
    l_space_inputs = tf.placeholder(tf.float32, (None, latent_space_z_dim), 'l_space_inputs')
    model_learning_rate = tf.placeholder(tf.float32, name='model_learning_rate')

    return true_inputs, l_space_inputs, model_learning_rate
```

2. 判别器

接下来，需要实现网络的判别器部分，该部分将用于判断输入是来自真实数据集还是由生成器生成的。同样，将使用 TensorFlow 的 tf.variable_scope 功能为一些变量添加带有判别器字样的前缀，以便可以检索和重用它们。

这里定义一个函数，它将返回判别器的二元输出和 logit 值。

```
# Defining the discriminator function
def discriminator(input_imgs, reuse=False):
    # using variable_scope to reuse variables
    with tf.variable_scope('discriminator', reuse=reuse):
        # leaky relu parameter
        leaky_param_alpha = 0.2

        # defining the layers
        conv_layer_1 = tf.layers.conv2d(input_imgs, 64, 5, 2, 'same')
        leaky_relu_output = tf.maximum(leaky_param_alpha * conv_layer_1, conv_layer_1)

        conv_layer_2 = tf.layers.conv2d(leaky_relu_output, 128, 5, 2, 'same')
        normalized_output = tf.layers.batch_normalization(conv_layer_2, training=True)
        leay_relu_output = tf.maximum(leaky_param_alpha *
```

```
            normalized_output, normalized_output)

            conv_layer_3 = tf.layers.conv2d(leay_relu_output, 256, 5, 2,
'same')
            normalized_output = tf.layers.batch_normalization(conv_layer_3,
training=True)
            leaky_relu_output = tf.maximum(leaky_param_alpha *
normalized_output, normalized_output)

            # reshaping the output for the logits to be 2D tensor
            flattened_output = tf.reshape(leaky_relu_output, (-1, 4 * 4 * 256))
            logits_layer = tf.layers.dense(flattened_output, 1)
            output = tf.sigmoid(logits_layer)

        return output, logits_layer
```

3. 生成器

本节将实现网络的第二部分,它尝试使用隐空间 z 复制原始输入图像。我们也将使用 tf.variable_scope 来实现此功能。

下面开始定义生成器函数,它将返回生成的图像。

```
def generator(z_latent_space, output_channel_dim, is_train=True):

    with tf.variable_scope('generator', reuse=not is_train):
        #leaky relu parameter
        leaky_param_alpha = 0.2
        fully_connected_layer = tf.layers.dense(z_latent_space, 2*2*512)
        #reshaping the output back to 4D tensor to match the accepted
format for convolution layer
        reshaped_output = tf.reshape(fully_connected_layer, (-1, 2, 2,
512))
        normalized_output = tf.layers.batch_normalization(reshaped_output,
training=is_train)
        leaky_relu_output = tf.maximum(leaky_param_alpha *
normalized_output, normalized_output)
        conv_layer_1 = tf.layers.conv2d_transpose(leaky_relu_output, 256,
5, 2, 'valid')
        normalized_output = tf.layers.batch_normalization(conv_layer_1,
training=is_train)
        leaky_relu_output = tf.maximum(leaky_param_alpha *
normalized_output, normalized_output)
        conv_layer_2 = tf.layers.conv2d_transpose(leaky_relu_output, 128,
```

```
5, 2, 'same')
        normalized_output = tf.layers.batch_normalization(conv_layer_2,
training=is_train)
        leaky_relu_output = tf.maximum(leaky_param_alpha *
normalized_output, normalized_output)
        logits_layer = tf.layers.conv2d_transpose(leaky_relu_output,
output_channel_dim, 5, 2, 'same')
        output = tf.tanh(logits_layer)
        return output
```

4. 模型的损失函数

现在出现了一个棘手的问题,这在第 14 章中也讨论过,那就是计算判别器和生成器的损失。

这里给出模型损失函数的定义,它将会利用先前定义的生成器和判别器函数。

```
# Define the error for the discriminator and generator
def model_losses(input_actual, input_latent_z, out_channel_dim):
    # building the generator part
    gen_model = generator(input_latent_z, out_channel_dim)
    disc_model_true, disc_logits_true = discriminator(input_actual)
    disc_model_fake, disc_logits_fake = discriminator(gen_model,
reuse=True)

    disc_loss_true = tf.reduce_mean(
        tf.nn.sigmoid_cross_entropy_with_logits(logits=disc_logits_true,
labels=tf.ones_like(disc_model_true)))

    disc_loss_fake =
tf.reduce_mean(tf.nn.sigmoid_cross_entropy_with_logits(
        logits=disc_logits_fake, labels=tf.zeros_like(disc_model_fake)))

    gen_loss = tf.reduce_mean(tf.nn.sigmoid_cross_entropy_with_logits(
        logits=disc_logits_fake, labels=tf.ones_like(disc_model_fake)))

    disc_loss = disc_loss_true + disc_loss_fake

    return disc_loss, gen_loss
```

5. 模型的优化器

最后,在训练模型之前,需要为此任务实现优化标准。这里将使用之前使用的命名约定来检索判别器和生成器的可训练参数并训练它们。

```
# specifying the optimization criteria
def model_optimizer(disc_loss, gen_loss, learning_rate, beta1):
    trainable_vars = tf.trainable_variables()
    disc_vars = [var for var in trainable_vars if
var.name.startswith('discriminator')]
    gen_vars = [var for var in trainable_vars if
var.name.startswith('generator')]

    disc_train_opt = tf.train.AdamOptimizer(
        learning_rate, beta1=beta1).minimize(disc_loss, var_list=disc_vars)

    update_operations = tf.get_collection(tf.GraphKeys.UPDATE_OPS)
    gen_updates = [opt for opt in update_operations if
opt.name.startswith('generator')]

    with tf.control_dependencies(gen_updates):
        gen_train_opt = tf.train.AdamOptimizer(
            learning_rate, beta1).minimize(gen_loss, var_list=gen_vars)

    return disc_train_opt, gen_train_opt
```

6. 模型的训练环节

下面开始训练模型,看看生成器如何能够在某种程度上通过生成非常接近原始 CelebA 数据集的图像来欺骗判别器。

首先,需要定义一个辅助函数,它将显示生成器生成的一些图像。

```
# define a function to visualize some generated images from the generator
def show_generator_output(sess, num_images, input_latent_z,
output_channel_dim, img_mode):
    cmap = None if img_mode == 'RGB' else 'gray'
    latent_space_z_dim = input_latent_z.get_shape().as_list()[-1]
    examples_z = np.random.uniform(-1, 1, size=[num_images,
latent_space_z_dim])

    examples = sess.run(
        generator(input_latent_z, output_channel_dim, False),
        feed_dict={input_latent_z: examples_z})

    images_grid = utils.images_square_grid(examples, img_mode)
    pyplot.imshow(images_grid, cmap=cmap)
    pyplot.show()
```

然后，使用之前定义的辅助函数来构建模型输入、损失和优化标准。这里将它们堆叠在一起，并基于 CelebA 数据集开始训练模型。

```
def model_train(num_epocs, train_batch_size, z_dim, learning_rate, beta1,
get_batches, input_data_shape, data_img_mode):
    _, image_width, image_height, image_channels = input_data_shape

    actual_input, z_input, leaningRate = inputs(
        image_width, image_height, image_channels, z_dim)

    disc_loss, gen_loss = model_losses(actual_input, z_input,
image_channels)

    disc_opt, gen_opt = model_optimizer(disc_loss, gen_loss, learning_rate,
beta1)

    steps = 0
    print_every = 50
    show_every = 100
    model_loss = []
    num_images = 25

    with tf.Session() as sess:

        # initializing all the variables
        sess.run(tf.global_variables_initializer())

        for epoch_i in range(num_epocs):
            for batch_images in get_batches(train_batch_size):

                steps += 1
                batch_images *= 2.0
                z_sample = np.random.uniform(-1, 1, (train_batch_size,
z_dim))
                _ = sess.run(disc_opt, feed_dict={
                    actual_input: batch_images, z_input: z_sample,
leaningRate: learning_rate})
                _ = sess.run(gen_opt, feed_dict={
                    z_input: z_sample, leaningRate: learning_rate})

                if steps % print_every == 0:
```

```
                        train_loss_disc = disc_loss.eval({z_input: z_sample,
actual_input: batch_images})
                        train_loss_gen = gen_loss.eval({z_input: z_sample})

                    print("Epoch {}/{}...".format(epoch_i + 1, num_epocs),
                          "Discriminator Loss:
{:.4f}...".format(train_loss_disc),
                          "Generator Loss: {:.4f}".format(train_loss_gen))
                    model_loss.append((train_loss_disc, train_loss_gen))

                if steps % show_every == 0:
                    show_generator_output(sess, num_images, z_input,
image_channels, data_img_mode)
```

开始训练,根据主机的配置,训练可能需要一些时间。

```
# Training the model on CelebA dataset
train_batch_size = 64
z_dim = 100
learning_rate = 0.002
beta1 = 0.5

num_epochs = 1

celeba_dataset = utils.Dataset('celeba', glob(os.path.join(data_dir,
'img_align_celeba/*.jpg')))
with tf.Graph().as_default():
    model_train(num_epochs, train_batch_size, z_dim, learning_rate, beta1,
celeba_dataset.get_batches,
                celeba_dataset.shape, celeba_dataset.image_mode)
```

输出如下。

```
Epoch 1/1... Discriminator Loss: 0.9118... Generator Loss: 12.2238
Epoch 1/1... Discriminator Loss: 0.6119... Generator Loss: 3.2168
Epoch 1/1... Discriminator Loss: 0.5383... Generator Loss: 2.8054
Epoch 1/1... Discriminator Loss: 1.4381... Generator Loss: 0.4672
Epoch 1/1... Discriminator Loss: 0.7815... Generator Loss: 14.8220
Epoch 1/1... Discriminator Loss: 0.6435... Generator Loss: 9.2591
Epoch 1/1... Discriminator Loss: 1.5661... Generator Loss: 10.4747
Epoch 1/1... Discriminator Loss: 1.5407... Generator Loss: 0.5811
Epoch 1/1... Discriminator Loss: 0.6470... Generator Loss: 2.9002
```

```
Epoch 1/1... Discriminator Loss: 0.5671... Generator Loss: 2.0700
Epoch 1/1... Discriminator Loss: 0.7950... Generator Loss: 1.5818
Epoch 1/1... Discriminator Loss: 1.2417... Generator Loss: 0.7094
Epoch 1/1... Discriminator Loss: 1.1786... Generator Loss: 1.0948
Epoch 1/1... Discriminator Loss: 1.0427... Generator Loss: 2.8878
Epoch 1/1... Discriminator Loss: 0.8409... Generator Loss: 2.6785
Epoch 1/1... Discriminator Loss: 0.8557... Generator Loss: 1.7706
Epoch 1/1... Discriminator Loss: 0.8241... Generator Loss: 1.2898
Epoch 1/1... Discriminator Loss: 0.8590... Generator Loss: 1.8217
Epoch 1/1... Discriminator Loss: 1.1694... Generator Loss: 0.8490
Epoch 1/1... Discriminator Loss: 0.9984... Generator Loss: 1.0042
```

训练前期与训练中期生成的样本分别如图 15.3 和图 15.4 所示。

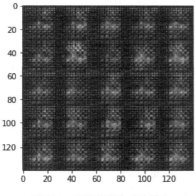

图 15.3　训练前期生成的样本

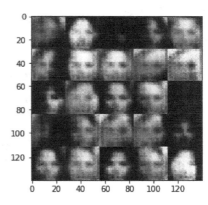

图 15.4　训练中期生成的样本

经过一段时间的训练,我们得到图 15.5 所示的样本。

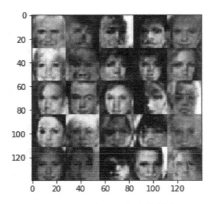

图 15.5　训练后期生成的样本

15.2 用生成对抗网络进行半监督学习

读者需注意，半监督学习是一种技术，其中标记和未标记的数据都用于训练分类器。

这类分类器使用一小部分标记的数据和大量未标记的数据（来自同一领域）。目的是组合这些数据源来训练一个**深度卷积神经网络**（Deep Convolution Neural Network，DCNN）用来学习一个推断函数，以便将新数据点映射到期望结果。

这里提出了一个 GAN 模型，使用一个非常小的标记训练集对街景房屋号码进行分类。事实上，模型大约用了原有的 SVHN 训练标记的 1.3%，即 1000 个标记样本。这里用了论文"Improved Techniques for Training GANs from OpenAI"（见 arxiv 网站）中提到的一些技术。

15.2.1 直观解释

在构建一个用于生成图像的 GAN 时，要同时训练生成器和判别器，如图 15.6 所示。在训练之后，将丢弃判别器，因为开发人员只用它来训练生成器。

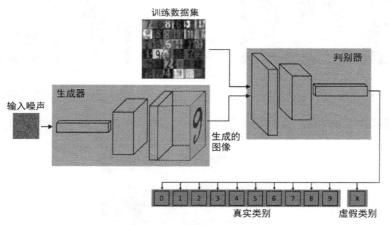

图 15.6 用在一个 11 类别分类问题中的半监督学习 GAN 架构

在半监督学习中，需要将判别器转换为多类别分类器。即使并没有许多的标记样本用来训练，这个新模型也必须能够在测试集上很好地泛化。另外，这里训练结束后，开发人员实际上可以丢掉生成器。要注意这里的角色变化。现在，生成器仅仅是用来在训练期间帮助判别器的。换句话说，生成器充当了一个不同的信息源，判别器从中获得原始的、未标记的训练数据。正如我们将看到的，这些未标记的数据是提升判别器性能的关键。同时，对于常规的图像生成 GAN，判别器只有一个角色。计算输入是否是真实图像的概率，这称为 GAN 问题。

然而，为了把判别器变成半监督学习中的分类器，除了 GAN 问题之外，判别器还需要学习原始数据集中每个类别的概率。也就是说，对于每幅输入图像，判别器需要学习它是一、二、三等类别的概率。

在图像生成的 GAN 判别器中，有单个 sigmoid 单元输出。这个数值表示了输入图像是真（数值接近于 1）或假（数值接近于 0）的概率。换句话说，从判别器的角度看，数值接近 1 表示该样本很有可能来自训练集，数值接近 0 表示样本很有可能是从生成网络来的。运用这个概率，判别器能够向生成器输送一个反馈信号。该信号使生成器能够在训练过程中调整参数，并提升生成器构建真实图像的能力。

我们需要把判别器（来自之前的 GAN）转换为 11 类别的分类器。为此，可以把它的 sigmoid 输出变成具有 11 个输出的 softmax，前 10 个类别是 SVHN 数据集中各个类别的概率（0~9），而第 11 个类别是所有从生成器中产生的虚假图像。

 如果把第 11 个类别的概率设为 0，那么前 10 个概率的和表示的是与 sigmoid 函数计算结果一样的概率。

最后，需要设置损失函数，使得判别器能够做到以下两点。

- 帮助生成器产生真实图像。为此，需要引导判别器区分真实和虚假样本。
- 使用生成器的图像，以及标记和未标记的训练数据，来帮助对数据集进行分类。

总的来说，判别器有 3 类不同来源的训练数据。

- 带标记的真实图像：这是图像标签对，和任何常规的监督分类问题一样。
- 不带标记的真实图像：对于这些图像，分类器只能学到这些图像是真的。
- 从生成器产生的图像：对于这些数据，判别器学习分辨它们是假的。

这些不同数据源的组合将使分类器能够从更广泛的角度进行学习。反之，这也使得模型能够比仅使用 1000 个带标记的样本进行训练时更精确地进行推理。

15.2.2 数据分析与预处理

数据分析与预处理将使用 SVHN 数据集，SVHV 是斯坦福的街景房屋号码（Street View House Numbers）的缩写。现在从导入这一实现所需的类库开始。

```python
# Lets start by loading the necessary libraries
%matplotlib inline

import pickle as pkl
import time
import matplotlib.pyplot as plt
import numpy as np
from scipy.io import loadmat
import tensorflow as tf
import os
```

接下来,定义一个辅助类来下载SVHN数据集(记住,要先手动创建input_data_dir)。

```python
from urllib.request import urlretrieve
from os.path import isfile, isdir
from tqdm import tqdm
input_data_dir = 'input/'

input_data_dir = 'input/'

if not isdir(input_data_dir):
    raise Exception("Data directory doesn't exist!")

class DLProgress(tqdm):
    last_block = 0

    def hook(self, block_num=1, block_size=1, total_size=None):
        self.total = total_size
        self.update((block_num - self.last_block) * block_size)
        self.last_block = block_num

if not isfile(input_data_dir + "train_32x32.mat"):
    with DLProgress(unit='B', unit_scale=True, miniters=1, desc='SVHN Training Set') as pbar:
        urlretrieve(
            'http://ufldl.stanford.edu/housenumbers/train_32x32.mat',
            input_data_dir + 'train_32x32.mat',
            pbar.hook)

if not isfile(input_data_dir + "test_32x32.mat"):
    with DLProgress(unit='B', unit_scale=True, miniters=1, desc='SVHN
```

```
Training Set') as pbar:
        urlretrieve(
            'http://ufldl.stanford.edu/housenumbers/test_32x32.mat',
            input_data_dir + 'test_32x32.mat',
            pbar.hook)

train_data = loadmat(input_data_dir + 'train_32x32.mat')
test_data = loadmat(input_data_dir + 'test_32x32.mat')
```

输出如下。

```
trainset shape: (32, 32, 3, 73257)
testset shape: (32, 32, 3, 26032)
```

通过以下代码看一看这些图像的样子。

```
indices = np.random.randint(0, train_data['X'].shape[3], size=36)
fig, axes = plt.subplots(6, 6, sharex=True, sharey=True, figsize=(5,5),)
for ii, ax in zip(indices, axes.flatten()):
    ax.imshow(train_data['X'][:,:,:,ii], aspect='equal')
    ax.xaxis.set_visible(False)
    ax.yaxis.set_visible(False)
plt.subplots_adjust(wspace=0, hspace=0)
```

输出如图 15.7 所示。

图 15.7 从 SVHN 数据集中的采样图像

接下来，我们需要把图像缩放到 -1~1，这个操作很有必要，因为会用到 tanh() 函数来压缩生成器的输出值。

```python
# Scaling the input images
def scale_images(image, feature_range=(-1, 1)):
    # scale image to (0, 1)
    image = ((image - image.min()) / (255 - image.min()))

    # scale the image to feature range
    min, max = feature_range
    image = image * (max - min) + min
    return image

class Dataset:
    def __init__(self, train_set, test_set, validation_frac=0.5, shuffle_data=True, scale_func=None):
        split_ind = int(len(test_set['y']) * (1 - validation_frac))
        self.test_input, self.valid_input = test_set['X'][:, :, :, :split_ind], test_set['X'][:, :, :, split_ind:]
        self.test_target, self.valid_target = test_set['y'][:split_ind], test_set['y'][split_ind:]
        self.train_input, self.train_target = train_set['X'], train_set['y']

        # The street house number dataset comes with lots of labels,
        # but because we are going to do semi-supervised learning we are going to assume that we don't have all labels
        # like, assume that we have only 1000
        self.label_mask = np.zeros_like(self.train_target)
        self.label_mask[0:1000] = 1

        self.train_input = np.rollaxis(self.train_input, 3)
        self.valid_input = np.rollaxis(self.valid_input, 3)
        self.test_input = np.rollaxis(self.test_input, 3)

        if scale_func is None:
            self.scaler = scale_images
        else:
            self.scaler = scale_func
        self.train_input = self.scaler(self.train_input)
        self.valid_input = self.scaler(self.valid_input)
        self.test_input = self.scaler(self.test_input)
        self.shuffle = shuffle_data

    def batches(self, batch_size, which_set="train"):
```

```
            input_name = which_set + "_input"
            target_name = which_set + "_target"

            num_samples = len(getattr(dataset, target_name))
            if self.shuffle:
                indices = np.arange(num_samples)
                np.random.shuffle(indices)
                setattr(dataset, input_name, getattr(dataset,
input_name)[indices])
                setattr(dataset, target_name, getattr(dataset,
target_name)[indices])
                if which_set == "train":
                    dataset.label_mask = dataset.label_mask[indices]

            dataset_input = getattr(dataset, input_name)
            dataset_target = getattr(dataset, target_name)

            for jj in range(0, num_samples, batch_size):
                input_vals = dataset_input[jj:jj + batch_size]
                target_vals = dataset_target[jj:jj + batch_size]

                if which_set == "train":
                    # including the label mask in case of training
                    # to pretend that we don't have all the labels

                    yield input_vals, target_vals, self.label_mask[jj:jj +
batch_size]
                else:
                    yield input_vals, target_vals
```

15.2.3 构建模型

在本节中,我们将构建测试所需要的细节,因此下面从定义用来输送数据到计算图的输入开始。

1. 模型输入

首先,定义模型的输入函数,构建用于输送数据到计算图的输入占位符。

```
# defining the model inputs
def inputs(actual_dim, z_dim):
    inputs_actual = tf.placeholder(tf.float32, (None, *actual_dim),
```

```
name='input_actual')
    inputs_latent_z = tf.placeholder(tf.float32, (None, z_dim),
name='input_latent_z')

    target = tf.placeholder(tf.int32, (None), name='target')
    label_mask = tf.placeholder(tf.int32, (None), name='label_mask')

    return inputs_actual, inputs_latent_z, target, label_mask
```

2. 生成器

下面实现 GAN 的第一个核心部分,这部分的架构和实现将以原本的 DCGAN 文章为依据。

```
def generator(latent_z, output_image_dim, reuse_vars=False,
leaky_alpha=0.2, is_training=True, size_mult=128):
    with tf.variable_scope('generator', reuse=reuse_vars):
        # define a fully connected layer
        fully_conntected_1 = tf.layers.dense(latent_z, 4 * 4 * size_mult * 4)

        # Reshape it from 2D tensor to 4D tensor to be fed to the convolution neural network
        reshaped_out_1 = tf.reshape(fully_conntected_1, (-1, 4, 4, size_mult * 4))
        batch_normalization_1 = tf.layers.batch_normalization(reshaped_out_1, training=is_training)
        leaky_output_1 = tf.maximum(leaky_alpha * batch_normalization_1, batch_normalization_1)

        conv_layer_1 = tf.layers.conv2d_transpose(leaky_output_1, size_mult * 2, 5, strides=2, padding='same')
        batch_normalization_2 = tf.layers.batch_normalization(conv_layer_1, training=is_training)
        leaky_output_2 = tf.maximum(leaky_alpha * batch_normalization_2, batch_normalization_2)

        conv_layer_2 = tf.layers.conv2d_transpose(leaky_output_2, size_mult, 5, strides=2, padding='same')
        batch_normalization_3 = tf.layers.batch_normalization(conv_layer_2, training=is_training)
        leaky_output_3 = tf.maximum(leaky_alpha * batch_normalization_
```

```
3, batch_normalization_3)

        # defining the output layer
        logits_layer = tf.layers.conv2d_transpose(leaky_output_3,
output_image_dim, 5, strides=2, padding='same')

        output = tf.tanh(logits_layer)

        return output
```

3. 判别器

下面构建 GAN 的第二个核心部分，即判别器。在之前的实现中，判别器将产生一个二值输出，表示输入数据是从真实数据来的（1）还是从生成器中生成的（0）。因为这里的场景有所不同，所以判别器将是一个多类别分类器。

下面继续构建架构中的判别器部分。

```
# Defining the discriminator part of the network
def discriminator(input_x, reuse_vars=False, leaky_alpha=0.2,
drop_out_rate=0., num_classes=10, size_mult=64):
    with tf.variable_scope('discriminator', reuse=reuse_vars):

        # defining a dropout layer
        drop_out_output = tf.layers.dropout(input_x, rate=drop_out_rate /
2.5)
        # Defining the input layer for the discriminator which is 32x32x3
        conv_layer_3 = tf.layers.conv2d(input_x, size_mult, 3, strides=2,
padding='same')
        leaky_output_4 = tf.maximum(leaky_alpha * conv_layer_3, conv_
layer_3)
        leaky_output_4 = tf.layers.dropout(leaky_output_4, rate=drop_
out_rate)

        conv_layer_4 = tf.layers.conv2d(leaky_output_4, size_mult, 3,
strides=2, padding='same')
        batch_normalization_4 = tf.layers.batch_normalization(conv_
layer_4, training=True)
        leaky_output_5 = tf.maximum(leaky_alpha * batch_normalization_
4, batch_normalization_4)

        conv_layer_5 = tf.layers.conv2d(leaky_output_5, size_mult, 3,
```

```
strides=2, padding='same')
        batch_normalization_5 = tf.layers.batch_normalization(conv_
layer_5, training=True)
        leaky_output_6 = tf.maximum(leaky_alpha * batch_normalization_
5, batch_normalization_5)
        leaky_output_6 = tf.layers.dropout(leaky_output_6, rate=drop_
out_rate)

        conv_layer_6 = tf.layers.conv2d(leaky_output_6, 2 * size_mult,
3, strides=1, padding='same')
        batch_normalization_6 = tf.layers.batch_normalization(conv_
layer_6, training=True)
        leaky_output_7 = tf.maximum(leaky_alpha * batch_normalization_
6, batch_normalization_6)

        conv_layer_7 = tf.layers.conv2d(leaky_output_7, 2 * size_mult,
3, strides=1, padding='same')
        batch_normalization_7 = tf.layers.batch_normalization(conv_
layer_7, training=True)
        leaky_output_8 = tf.maximum(leaky_alpha * batch_normalization_7,
batch_normalization_7)

        conv_layer_8 = tf.layers.conv2d(leaky_output_8, 2 * size_mult,
3, strides=2, padding='same')
        batch_normalization_8 = tf.layers.batch_normalization(conv_
layer_8, training=True)
        leaky_output_9 = tf.maximum(leaky_alpha * batch_normalization_
8, batch_normalization_8)
        leaky_output_9 = tf.layers.dropout(leaky_output_9, rate=drop_
out_rate)

        conv_layer_9 = tf.layers.conv2d(leaky_output_9, 2 * size_mult,
3, strides=1, padding='valid')

        leaky_output_10 = tf.maximum(leaky_alpha * conv_layer_9, conv_
layer_9)
        ...
```

这里没有在结尾使用一个全连接层，而是采用称为**全局平均池化**（global average pooling，GAP）的层，它取特征向量空间维度的平均值，这将产生仅有一个单值的压缩张量。

15.2 用生成对抗网络进行半监督学习

```
...
# Flatten it by global average pooling
leaky_output_features = tf.reduce_mean(leaky_output_10, (1, 2))
...
```

例如，假设在很多卷积之后，得到了一个输出张量，形状如下。

`[BATCH_SIZE, 8, 8, NUM_CHANNELS]`

为了使用全局平均池化，需要计算[8×8]张量片段的平均值。该操作将得到一个如下形状的张量。

`[BATCH_SIZE, 1, 1, NUM_CHANNELS]`

可以把它重塑为：

`[BATCH_SIZE, NUM_CHANNELS]`

在使用全局平均池化之后，加上一个全连接层，它将输出最后的对数，其形状如下。

`[BATCH_SIZE, NUM_CLASSES]`

它表示各个类别的分数。为了得到这些分数的概率，将使用 softmax 激活函数。

```
...
# Get the probability that the input is real rather than fake
softmax_output = tf.nn.softmax(classes_logits)s
...
```

最后的判别器函数如下。

```
# Defining the discriminator part of the network
def discriminator(input_x, reuse_vars=False, leaky_alpha=0.2,
drop_out_rate=0., num_classes=10, size_mult=64):
    with tf.variable_scope('discriminator', reuse=reuse_vars):
        
        # defining a dropout layer
        drop_out_output = tf.layers.dropout(input_x, rate=drop_out_rate / 2.5)
        
        # Defining the input layer for the discrminator which is 32x32x3
```

```
        conv_layer_3 = tf.layers.conv2d(input_x, size_mult, 3, strides=2,
padding='same')
        leaky_output_4 = tf.maximum(leaky_alpha * conv_layer_3, conv_
layer_3)
        leaky_output_4 = tf.layers.dropout(leaky_output_4, rate=drop_
out_rate)

        conv_layer_4 = tf.layers.conv2d(leaky_output_4, size_mult, 3,
strides=2, padding='same')
        batch_normalization_4 = tf.layers.batch_normalization(conv_
layer_4, training=True)
        leaky_output_5 = tf.maximum(leaky_alpha * batch_normalization_
4, batch_normalization_4)

        conv_layer_5 = tf.layers.conv2d(leaky_output_5, size_mult, 3,
strides=2, padding='same')
        batch_normalization_5 = tf.layers.batch_normalization(conv_
layer_5, training=True)
        leaky_output_6 = tf.maximum(leaky_alpha * batch_normalization_
5, batch_normalization_5)
        leaky_output_6 = tf.layers.dropout(leaky_output_6, rate=drop_
out_rate)

        conv_layer_6 = tf.layers.conv2d(leaky_output_6, 2 * size_mult,
3, strides=1, padding='same')
        batch_normalization_6 = tf.layers.batch_normalization(conv_
layer_6, training=True)
        leaky_output_7 = tf.maximum(leaky_alpha * batch_normalization_
6, batch_normalization_6)

        conv_layer_7 = tf.layers.conv2d(leaky_output_7, 2 * size_mult,
3, strides=1, padding='same')
        batch_normalization_7 = tf.layers.batch_normalization(conv_
layer_7, training=True)
        leaky_output_8 = tf.maximum(leaky_alpha * batch_normalization_
7, batch_normalization_7)

        conv_layer_8 = tf.layers.conv2d(leaky_output_8, 2 * size_mult,
3, strides=2, padding='same')
        batch_normalization_8 = tf.layers.batch_normalization(conv_
layer_8, training=True)
        leaky_output_9 = tf.maximum(leaky_alpha * batch_normalization_
8, batch_normalization_8)
```

```python
        leaky_output_9 = tf.layers.dropout(leaky_output_9, rate=drop_out_rate)

        conv_layer_9 = tf.layers.conv2d(leaky_output_9, 2 * size_mult, 3, strides=1, padding='valid')

        leaky_output_10 = tf.maximum(leaky_alpha * conv_layer_9, conv_layer_9)

        # Flatten it by global average pooling
        leaky_output_features = tf.reduce_mean(leaky_output_10, (1, 2))

        # Set class_logits to be the inputs to a softmax distribution over the different classes
        classes_logits = tf.layers.dense(leaky_output_features, num_classes + extra_class)

        if extra_class:
            actual_class_logits, fake_class_logits = tf.split(classes_logits, [num_classes, 1], 1)
            assert fake_class_logits.get_shape()[1] == 1, fake_class_logits.get_shape()
            fake_class_logits = tf.squeeze(fake_class_logits)
        else:
            actual_class_logits = classes_logits
            fake_class_logits = 0.

        max_reduced = tf.reduce_max(actual_class_logits, 1, keep_dims=True)
        stable_actual_class_logits = actual_class_logits - max_reduced

        gan_logits = tf.log(tf.reduce_sum(tf.exp(stable_actual_class_logits), 1)) + tf.squeeze(max_reduced) - fake_class_logits

        softmax_output = tf.nn.softmax(classes_logits)

        return softmax_output, classes_logits, gan_logits, leaky_output_features
```

4. 模型的损失函数

下面定义模型的损失。判别器损失分为两部分：第一部分表示 GAN 问题，即无监督损失；第二部分计算各个真实类别的概率，即监督损失。

对于判别器的无监督损失，需要区分真实训练图像和生成器生成的图像。

对一个常规的 GAN 来说，在一半的时间里判别器会以从训练集得到的未标记图像作为输入，在另一半的时间里以从生成器中生成的虚假、未标记的图像作为输入。

对于判别器损失的第二部分，即监督损失，需要基于判别器的对数来构建。因为这是一个多分类问题，所以将用到 softmax 交叉熵。

正如"Enhanced Techniques for Training GANs"一文中提到的，应该对生成器的损失使用特征匹配。作者的描述为：

"特征匹配的概念是惩罚训练数据上的某些特征集的平均值与生成样本上的那个特征集的平均值的平均绝对误差。为此，要从两个不同的来源中获取一些统计数据（矩）并强制它们相似。首先，获取在处理真实训练的最小批数据时从判别器中提取的特征的平均值。其次，当判别器在分析包含从生成器产生的虚假图像的最小批时，用相同的方式计算矩。最后，在这两个矩集合的基础上，生成器的损失就是它们之间的平均绝对差。换句话说，如文章所强调的：训练生成器以匹配判别器中间层上的特征的预期值。"

最终，模型的损失函数如下。

```
def model_losses(input_actual, input_latent_z, output_dim, target,
num_classes, label_mask, leaky_alpha=0.2,
                    drop_out_rate=0.):

    # These numbers multiply the size of each layer of the generator
and the discriminator,

    # respectively. You can reduce them to run your code faster for
debugging purposes.
    gen_size_mult = 32
    disc_size_mult = 64

    # Here we run the generator and the discriminator
    gen_model = generator(input_latent_z, output_dim, leaky_alpha=
leaky_alpha, size_mult=gen_size_mult)
    disc_on_data = discriminator(input_actual, leaky_alpha=leaky_
alpha, drop_out_rate=drop_out_rate,
                                 size_mult=disc_size_mult)
    disc_model_real, class_logits_on_data, gan_logits_on_data, data_
```

```
        features = disc_on_data
        disc_on_samples = discriminator(gen_model, reuse_vars=True,
leaky_alpha=leaky_alpha,
                                        drop_out_rate=drop_out_rate,
size_mult=disc_size_mult)
        disc_model_fake, class_logits_on_samples, gan_logits_on_samples,
sample_features = disc_on_samples

        # Here we compute `disc_loss`, the loss for the discriminator.
        disc_loss_actual = tf.reduce_mean(tf.nn.sigmoid_cross_entropy_
with_logits(logits=gan_logits_on_data, labels=tf.ones_like(gan_logits_
on_data)))
        disc_loss_fake = tf.reduce_mean(tf.nn.sigmoid_cross_entropy_with_
logits(logits=gan_logits_on_samples, labels=tf.zeros_like(gan_logits_
on_samples)))
        target = tf.squeeze(target)
        classes_cross_entropy = tf.nn.softmax_cross_entropy_with_logits
(logits=class_logits_on_data, labels=tf.one_hot(target, num_classes +
extra_class, dtype=tf.float32))
        classes_cross_entropy = tf.squeeze(classes_cross_entropy)
        label_m = tf.squeeze(tf.to_float(label_mask))
        disc_loss_class = tf.reduce_sum(label_m * classes_cross_entropy) /
tf.maximum(1., tf.reduce_sum(label_m))
        disc_loss = disc_loss_class + disc_loss_actual + disc_loss_fake

        # Here we set `gen_loss` to the "feature matching" loss invented
by Tim Salimans.
        sampleMoments = tf.reduce_mean(sample_features, axis=0)
        dataMoments = tf.reduce_mean(data_features, axis=0)

        gen_loss = tf.reduce_mean(tf.abs(dataMoments - sampleMoments))

        prediction_class = tf.cast(tf.argmax(class_logits_on_data, 1),
tf.int32)

        check_prediction = tf.equal(tf.squeeze(target), prediction_class)
        correct = tf.reduce_sum(tf.to_float(check_prediction))
        masked_correct = tf.reduce_sum(label_m * tf.to_float(check_
prediction))

        return disc_loss, gen_loss, correct, masked_correct, gen_model
```

5. 模型的优化器

下面定义模型的优化器，它和之前定义的那个优化器十分类似。

```
def model_optimizer(disc_loss, gen_loss, learning_rate, beta1):

    # Get weights and biases to update. Get them separately for the discriminator and the generator
    trainable_vars = tf.trainable_variables()
    disc_vars = [var for var in trainable_vars if var.name.startswith('discriminator')]
    gen_vars = [var for var in trainable_vars if var.name.startswith('generator')]
    for t in trainable_vars:
        assert t in disc_vars or t in gen_vars

    # Minimize both gen and disc costs simultaneously
    disc_train_optimizer = tf.train.AdamOptimizer(learning_rate, beta1=beta1).minimize(disc_loss, var_list=disc_vars)
    gen_train_optimizer = tf.train.AdamOptimizer(learning_rate, beta1=beta1).minimize(gen_loss, var_list=gen_vars)
    shrink_learning_rate = tf.assign(learning_rate, learning_rate * 0.9)

    return disc_train_optimizer, gen_train_optimizer, shrink_learning_rate
```

6. 模型的训练环节

最后，将上述所有部分合在一起开始进入训练过程。

```
class GAN:
    def __init__(self, real_size, z_size, learning_rate, num_classes=10, alpha=0.2, beta1=0.5):
        tf.reset_default_graph()

        self.learning_rate = tf.Variable(learning_rate, trainable=False)
        model_inputs = inputs(real_size, z_size)
        self.input_actual, self.input_latent_z, self.target, self.label_mask = model_inputs
        self.drop_out_rate = tf.placeholder_with_default(.5, (),
```

```python
                "drop_out_rate")

            losses_results = model_losses(self.input_actual, self.input_
latent_z,
                                    real_size[2], self.target,
num_classes,
                                    label_mask=self.label_mask,
                                    leaky_alpha=0.2,
                                    drop_out_rate=self.drop_out_rate)
            self.disc_loss, self.gen_loss, self.correct, self.masked_
correct, self.samples = losses_results

            self.disc_opt, self.gen_opt, self.shrink_learning_rate =
model_optimizer(self.disc_loss, self.gen_loss, self.learning_rate, beta1)

def view_generated_samples(epoch, samples, nrows, ncols, figsize=(5, 5)):
        fig, axes = plt.subplots(figsize=figsize, nrows=nrows, ncols=ncols,
                            sharey=True, sharex=True)
        for ax, img in zip(axes.flatten(), samples[epoch]):
            ax.axis('off')
            img = ((img - img.min()) * 255 / (img.max() - img.min())).
astype(np.uint8)
            ax.set_adjustable('box-forced')
            im = ax.imshow(img)

        plt.subplots_adjust(wspace=0, hspace=0)
        return fig, axes

def train(net, dataset, epochs, batch_size, figsize=(5, 5)):

        saver = tf.train.Saver()
        sample_z = np.random.normal(0, 1, size=(50, latent_space_z_size))

        samples, train_accuracies, test_accuracies = [], [], []
        steps = 0

        with tf.Session() as sess:
            sess.run(tf.global_variables_initializer())
            for e in range(epochs):
                print("Epoch", e)

                num_samples = 0
```

```python
            num_correct_samples = 0
            for x, y, label_mask in dataset.batches(batch_size):
                assert 'int' in str(y.dtype)
                steps += 1
                num_samples += label_mask.sum()

                # Sample random noise for G
                batch_z = np.random.normal(0, 1, size=(batch_size,
latent_space_z_size))

                _, _, correct = sess.run([net.disc_opt, net.gen_opt,
net.masked_correct],
                                         feed_dict={net.input_actual:
x, net.input_latent_z: batch_z,
                                                    net.target: y,
net.label_mask: label_mask})
                num_correct_samples += correct

            sess.run([net.shrink_learning_rate])

            training_accuracy = num_correct_samples /
float(num_samples)

            print("\t\tClassifier train accuracy: ", training_accuracy)

            num_samples = 0
            num_correct_samples = 0

            for x, y in dataset.batches(batch_size, which_set="test"):
                assert 'int' in str(y.dtype)
                num_samples += x.shape[0]

                correct, = sess.run([net.correct],
feed_dict={net.input_real: x,
                                                    net.y: y,
net.drop_rate: 0.})
                num_correct_samples += correct

            testing_accuracy = num_correct_samples / float(num_samples)
            print("\t\tClassifier test accuracy", testing_accuracy)

            gen_samples = sess.run(
                net.samples,
```

```
                    feed_dict={net.input_latent_z: sample_z})
            samples.append(gen_samples)
            _ = view_generated_samples(-1, samples, 5, 10,
figsize=figsize)
            plt.show()

            # Save history of accuracies to view after training
            train_accuracies.append(training_accuracy)
            test_accuracies.append(testing_accuracy)

        saver.save(sess, './checkpoints/generator.ckpt')

    with open('samples.pkl', 'wb') as f:
        pkl.dump(samples, f)

    return train_accuracies, test_accuracies, samples
```

不要忘了创建检查点的目录。

```
real_size = (32,32,3)
latent_space_z_size = 100
learning_rate = 0.0003

net = GAN(real_size, latent_space_z_size, learning_rate)

dataset = Dataset(train_data, test_data)

train_batch_size = 128
num_epochs = 25
train_accuracies, test_accuracies, samples = train(net,
                                                    dataset,
                                                    num_epochs,
                                                    train_batch_size,
                                                    figsize=(10,5))
```

最后,在 Epoch 24 中,我们获得如下输出,输出的样本图像如图 15.8 所示。

```
Epoch 24
            Classifier train accuracy:  0.937
            Classifier test accuracy 0.67401659496
            Step time:  0.03694915771484375
            Epoch time:  26.15842580795288
```

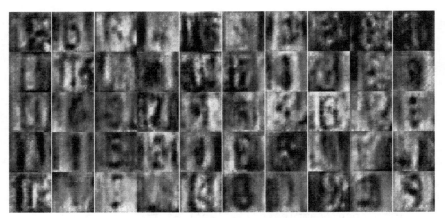

图 15.8 使用特征匹配损失的生成网络产生的样本图像

运行以下代码得到的结果如图 15.9 所示。

```
fig, ax = plt.subplots()
plt.plot(train_accuracies, label='Train', alpha=0.5)
plt.plot(test_accuracies, label='Test', alpha=0.5)
plt.title("Accuracy")
plt.legend()
```

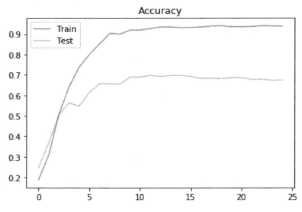

图 15.9 训练过程中的训练准确率与测试准确率

尽管特征匹配损失在半监督学习任务中表现良好,但是生成器产生的图像并不像之前的章节中产生的图像那样好。这个实现主要是为了演示如何使用 GAN 进行半监督学习的设定。

15.3 总结

许多研究人员认为无监督学习是一般 AI 系统中缺失的环节。为了克服这些障碍，尝试使用较少的标记数据来解决指定的问题是一个关键。在这种场景下，GAN 为使用少量标记的样本学习复杂的任务提供了另一个替代方案。然而，监督学习和半监督学习之间的性能差距仍然很大。随着新技术的产生，这种差距有望越来越小。

附录 A
实现鱼类识别

以下是第 1 章中鱼类识别部分的完整代码。

鱼类识别部分的代码

在讲解了鱼类识别示例的主要构建模块后，我们将所有代码段落连接在一起，来看看如何使用几行代码构建一个复杂的系统。

```python
#Loading the required libraries along with the deep learning platform Keras
with TensorFlow as backend
import numpy as np
np.random.seed(2017)
import os
import glob
import cv2
import pandas as pd
import time
import warnings
from sklearn.cross_validation import KFold
from keras.models import Sequential
from keras.layers.core import Dense, Dropout, Flatten
from keras.layers.convolutional import Convolution2D, MaxPooling2D, ZeroPadding2D
from keras.optimizers import SGD
from keras.callbacks import EarlyStopping
from keras.utils import np_utils
from sklearn.metrics import log_loss
from keras import __version__ as keras_version
```

```python
# Parameters
# ----------
# x : type
#    Description of parameter `x`.
def rezize_image(img_path):
    img = cv2.imread(img_path)
    img_resized = cv2.resize(img, (32, 32), cv2.INTER_LINEAR)
    return img_resized
#Loading the training samples from their corresponding folder names, where we have a folder for each type
def load_training_samples():
    #Variables to hold the training input and output variables
    train_input_variables = []
    train_input_variables_id = []
    train_label = []
    # Scanning all images in each folder of a fish type
    print('Start Reading Train Images')
    folders = ['ALB', 'BET', 'DOL', 'LAG', 'NoF', 'OTHER', 'SHARK', 'YFT']
    for fld in folders:
        folder_index = folders.index(fld)
        print('Load folder {} (Index: {})'.format(fld, folder_index))
        imgs_path = os.path.join('..', 'input', 'train', fld, '*.jpg')
        files = glob.glob(imgs_path)
        for file in files:
            file_base = os.path.basename(file)
            # Resize the image
            resized_img = rezize_image(file)
            # Appending the processed image to the input/output variables of the classifier
            train_input_variables.append(resized_img)
            train_input_variables_id.append(file_base)
            train_label.append(folder_index)
    return train_input_variables, train_input_variables_id, train_label
#Loading the testing samples which will be used to testing how well the model was trained
def load_testing_samples():
    # Scanning images from the test folder
    imgs_path = os.path.join('..', 'input', 'test_stg1', '*.jpg')
    files = sorted(glob.glob(imgs_path))
    # Variables to hold the testing samples
    testing_samples = []
    testing_samples_id = []
    #Processing the images and appending them to the array that we have
```

```python
    for file in files:
        file_base = os.path.basename(file)
        # Image resizing
        resized_img = rezize_image(file)
        testing_samples.append(resized_img)
        testing_samples_id.append(file_base)
    return testing_samples, testing_samples_id
# formatting the images to fit our model
def format_results_for_types(predictions, test_id, info):
    model_results = pd.DataFrame(predictions, columns=['ALB', 'BET', 'DOL', 'LAG', 'NoF', 'OTHER',
        'SHARK', 'YFT'])
    model_results.loc[:, 'image'] = pd.Series(test_id, index=model_results.index)
    sub_file = 'testOutput_' + info + '.csv'
    model_results.to_csv(sub_file, index=False)
def load_normalize_training_samples():
    # Calling the load function in order to load and resize the training samples
    training_samples, training_label, training_samples_id = load_training_samples()
    # Converting the loaded and resized data into Numpy format
    training_samples = np.array(training_samples, dtype=np.uint8)
    training_label = np.array(training_label, dtype=np.uint8)
    # Reshaping the training samples
    training_samples = training_samples.transpose((0, 3, 1, 2))
    # Converting the training samples and training labels into float format
    training_samples = training_samples.astype('float32')
    training_samples = training_samples / 255
    training_label = np_utils.to_categorical(training_label, 8)
    return training_samples, training_label, training_samples_id
#Loading and normalizing the testing sample to fit into our model
def load_normalize_testing_samples():
    # Calling the load function in order to load and resize the testing samples
    testing_samples, testing_samples_id = load_testing_samples()
    # Converting the loaded and resized data into Numpy format
    testing_samples = np.array(testing_samples, dtype=np.uint8)
    # Reshaping the testing samples
    testing_samples = testing_samples.transpose((0, 3, 1, 2))
    # Converting the testing samples into float format
    testing_samples = testing_samples.astype('float32')
    testing_samples = testing_samples / 255
```

```python
    return testing_samples, testing_samples_id
def merge_several_folds_mean(data, num_folds):
    a = np.array(data[0])
    for i in range(1, num_folds):
        a += np.array(data[i])
    a /= num_folds
    return a.tolist()
# Create CNN model architecture
def create_cnn_model_arch():
    pool_size = 2 # we will use 2x2 pooling throughout
    conv_depth_1 = 32 # we will initially have 32 kernels per conv. layer...
    conv_depth_2 = 64 # ...switching to 64 after the first pooling layer
    kernel_size = 3 # we will use 3x3 kernels throughout
    drop_prob = 0.5 # dropout in the FC layer with probability 0.5
    hidden_size = 32 # the FC layer will have 512 neurons
    num_classes = 8 # there are 8 fish types
    # Conv [32] -> Conv [32] -> Pool
    cnn_model = Sequential()
    cnn_model.add(ZeroPadding2D((1, 1), input_shape=(3, 32, 32),
dim_ordering='th'))
    cnn_model.add(Convolution2D(conv_depth_1, kernel_size, kernel_size,
activation='relu', dim_ordering='th'))
    cnn_model.add(ZeroPadding2D((1, 1), dim_ordering='th'))
    cnn_model.add(Convolution2D(conv_depth_1, kernel_size, kernel_size,
activation='relu', dim_ordering='th'))
    cnn_model.add(MaxPooling2D(pool_size=(pool_size, pool_size), strides=(2,
2), dim_ordering='th'))
    # Conv [64] -> Conv [64] -> Pool
    cnn_model.add(ZeroPadding2D((1, 1), dim_ordering='th'))
    cnn_model.add(Convolution2D(conv_depth_2, kernel_size, kernel_size,
activation='relu', dim_ordering='th'))
    cnn_model.add(ZeroPadding2D((1, 1), dim_ordering='th'))
    cnn_model.add(Convolution2D(conv_depth_2, kernel_size, kernel_size,
activation='relu', dim_ordering='th'))
    cnn_model.add(MaxPooling2D(pool_size=(pool_size, pool_size), strides=(2,
2), dim_ordering='th'))
    # Now flatten to 1D, apply FC then ReLU (with dropout) and finally
softmax(output layer)
    cnn_model.add(Flatten())
    cnn_model.add(Dense(hidden_size, activation='relu'))
    cnn_model.add(Dropout(drop_prob))
    cnn_model.add(Dense(hidden_size, activation='relu'))
    cnn_model.add(Dropout(drop_prob))
```

```python
    cnn_model.add(Dense(num_classes, activation='softmax'))
    # initiating the stochastic gradient descent optimiser
    stochastic_gradient_descent = SGD(lr=1e-2, decay=1e-6, momentum=0.9, nesterov=True)
    cnn_model.compile(optimizer=stochastic_gradient_descent, # using the stochastic gradient descent optimiser
                      loss='categorical_crossentropy') # using the cross-entropy loss function
    return cnn_model
#Model using with kfold cross validation as a validation method
def create_model_with_kfold_cross_validation(nfolds=10):
    batch_size = 16 # in each iteration, we consider 32 training examples at once
    num_epochs = 30 # we iterate 200 times over the entire training set
    random_state = 51 # control the randomness for reproducibility of the results on the same platform
    # Loading and normalizing the training samples prior to feeding it to the created CNN model
    training_samples, training_samples_target, training_samples_id = load_normalize_training_samples()
    yfull_train = dict()
    # Providing Training/Testing indices to split data in the training samples
    # which is splitting data into 10 consecutive folds with shuffling
    kf = KFold(len(train_id), n_folds=nfolds, shuffle=True, random_state=random_state)
    fold_number = 0 # Initial value for fold number
    sum_score = 0 # overall score (will be incremented at each iteration)
    trained_models = [] # storing the modeling of each iteration over the folds
    # Getting the training/testing samples based on the generated training/testing indices by Kfold
    for train_index, test_index in kf:
        cnn_model = create_cnn_model_arch()
        training_samples_X = training_samples[train_index] # Getting the training input variables
        training_samples_Y = training_samples_target[train_index] # Getting the training output/label variable
        validation_samples_X = training_samples[test_index] # Getting the validation input variables
        validation_samples_Y = training_samples_target[test_index] # Getting the validation output/label variabl
        fold_number += 1
```

```python
        print('Fold number {} out of {}'.format(fold_number, nfolds))
        callbacks = [
            EarlyStopping(monitor='val_loss', patience=3, verbose=0),
        ]
        # Fitting the CNN model giving the defined settings
        cnn_model.fit(training_samples_X, training_samples_Y, batch_size=batch_size,
            nb_epoch=num_epochs,
              shuffle=True, verbose=2, validation_data=(validation_samples_X,
                validation_samples_Y),
            callbacks=callbacks)
        # measuring the generalization ability of the trained model based on the validation set
        predictions_of_validation_samples = \
            cnn_model.predict(validation_samples_X.astype('float32'), batch_size=batch_size,
            verbose=2)
        current_model_score = log_loss(Y_valid, predictions_of_validation_samples)
        print('Current model score log_loss: ', current_model_score)
        sum_score += current_model_score*len(test_index)
        # Store valid predictions
        for i in range(len(test_index)):
            yfull_train[test_index[i]] = predictions_of_validation_samples[i]
        # Store the trained model
        trained_models.append(cnn_model)
    # incrementing the sum_score value by the current model calculated score
    overall_score = sum_score/len(training_samples)
    print("Log_loss train independent avg: ", overall_score)
    #Reporting the model loss at this stage
    overall_settings_output_string = 'loss_' + str(overall_score) + '_folds_'
+ str(nfolds) + '_ep_' + str(num_epochs)
    return overall_settings_output_string, trained_models
#Testing how well the model is trained
def test_generality_crossValidation_over_test_set(overall_settings_output_string, cnn_models):
    batch_size = 16 # in each iteration, we consider 32 training examples at once
    fold_number = 0 # fold iterator
    number_of_folds = len(cnn_models) # Creating number of folds based on the value used in the training step
    yfull_test = [] # variable to hold overall predictions for the test set
```

```
#executing the actual cross validation test process over the test set
for j in range(number_of_folds):
    model = cnn_models[j]
    fold_number += 1
    print('Fold number {} out of {}'.format(fold_number,
number_of_folds))
    #Loading and normalizing testing samples
    testing_samples, testing_samples_id =
load_normalize_testing_samples()
    #Calling the current model over the current test fold
    test_prediction = model.predict(testing_samples,
batch_size=batch_size, verbose=2)
    yfull_test.append(test_prediction)
test_result = merge_several_folds_mean(yfull_test, number_of_folds)
overall_settings_output_string = 'loss_' + overall_settings_output_string \
            + '_folds_' + str(number_of_folds)
format_results_for_types(test_result, testing_samples_id,
overall_settings_output_string)
# Start the model training and testing
if __name__ == '__main__':
    info_string, models = create_model_with_kfold_cross_validation()
    test_generality_crossValidation_over_test_set(info_string, models)
```